'A food revolution.'

***DAILY MAIL***

'Jessie Inchauspé, a born communicator, has written the best practical guide for managing glucose to maximise health and longevity.'

**DAVID SINCLAIR,** Professor of Genetics at Harvard Medical School and *New York Times* bestselling author of *Lifespan*

'*Glucose Revolution* will help you feel better, cut cravings, connect with yourself, balance your hormones, live longer, teach you science and put a smile on your face along the way.'

**DAVINIA TAYLOR,** actress and #1 *Sunday Times* bestselling author of *It's Not a Diet*

'A savvy, one-stop shop to sustainable weight loss and better health.'

**NINA TEICHOLZ,** *New York Times* bestselling author of *The Big Fat Surprise*

'Jessie Inchauspé takes the new science of nutrition and makes it practical for everyone.'

**ROBERT H. LUSTIG, MD, MSL,** author of *New York Times* bestseller *Fat Chance* and *Metabolical*

'Reading and applying the principles outlined in *Glucose Revolution* can help anyone achieve divine levels of health.'

**BENJAMIN BIKMAN, PHD,** Professor of Cell Biology, Brigham Young University, and author of *Why We Get Sick*

# TESTIMONIALS FROM THE GLUCOSE GODDESS INSTAGRAM COMMUNITY

*While these testimonials are based on individual success stories, results may vary.*

'A few days of applying Jessie's tips and my cravings disappeared. That changed everything.' —*Laura, 63*

'I'm eating pasta and losing weight. How much more awesome does it get?' —*Jasmin, 20*

'After two years of not ovulating, I'm regularly ovulating again. I lost 35lb. My acne cleared up. And mentally I am feeling so much better. The information Jessie shares changed my life. No going back!' —*Heather, 31*

'Jessie has shown me that I could change how menopause went for me. My friends told me I would never be able to lose the weight I had gained. Thanks to Jessie I proved them wrong! With her glucose hacks, I lost 9lb, I sleep like I used to, I feel amazing, and I no longer want to nap in the middle of the afternoon.' —*Bernadette, 55*

'I was diagnosed with type 2 diabetes after my third pregnancy 16 years ago. For years it's been getting worse and it's been difficult to manage. After starting to implement Jessie's hacks, in four months, I went from 11 mmol/L fasting glucose level to 6 mmol/L: from severely diabetic to diabetic no longer. I've been able to reverse the condition on my own!' —*Fatemeh, 51*

'Life-changing information... I lost 36lb in two months! My recurrent migraine problem improved significantly, and my energy is through the roof. I feel better than ever.' —*Annalaura, 49*

'In four months of following Jessie's glucose teachings, I've effortlessly lost 13lb, my massive hormonal acne is gone, and for the first time in my adult life I have normal thyroid levels (I went from 8.7 mIU/L TSH to 4.4 mIU/L). I have never felt better.' —*Tamara, 31*

'I'm a 64-year-old breast cancer survivor with heart, glucose and thyroid conditions. I take hormone suppressors and yet I have managed to lose 18lb in three months with the ridiculously easy changes Jessie explains so well. I'm the slimmest I've been since I gave birth and my blood tests are, in the words of my doctor, "those of a 15-year-old". It's hard to believe, even for me! Thank you, Jessie, for changing my life.' —*Dovra, 64*

'I'm a type 1 diabetic. I used to spike up to 16 mmol/L after breakfast. With the information Jessie shares I've learned to keep my glucose steady and my HbA1c has dropped from 7.4 per cent to 5.1 per cent in three months... I don't snap at my family and friends any more. I can finally be the person I want to be.' —*Lucy, 24*

'I have no words to describe how much Jessie's hacks changed my life. Two years ago I stopped taking the pill with the aim of starting a family. I thought it would be easy. But my period never came. After a year, I went to the doctor. I was diagnosed with insulin resistance and PCOS. It was really hard. Thankfully I found Jessie's work and I had hope again... I started implementing her tips. My period came back after two months! All my PCOS symptoms vanished (hair growth, anxiety, constant eating) and now... I have just found out I'm pregnant! I'm so happy I cannot describe it!' —*Filipa, 29*

'I went from 19 per cent body fat percentage to 8 per cent. So happy! And all this while eating all the foods I love.' — *Semir, 24*

'I was diagnosed with gestational diabetes at 29 weeks pregnant. So far, after one month of Jessie's tips, huge changes: I feel better than I ever did during the pregnancy, I'm not swollen, my blood sugar levels are steady and managed, my doctor is happy, and, most importantly, I'm not scared any more. Can't recommend Jessie's work enough for all moms-to-be.' —*Paulina, 39*

'I've been severely bulimic for nearly 30 years and nothing had ever helped until I started following Jessie and taking care of my glucose levels with her hacks. I haven't binged or purged for two months now, which is unbelievable. I honestly thought it was something that was just a part of me and I would never get over.' —*Sue, 48*

'I've dealt with hypoglycemia (low blood sugar) for a number of years. I was unaware that I could significantly improve it by just changing a few things about how I ate, like the order in which to eat my food. Thanks to Jessie and her evidence-based observations, I've learned how to eat a cookie or chocolate with far fewer negative impacts. Now that my blood sugar is more stable, I'm able to better address my anxiety symptoms and focus on dealing with its root causes.' —*Ilana, 37*

'In one month I feel like I've been reborn. I've had myalgic encephalomyelitis and chronic fatigue most of my life. I have also struggled with long Covid symptoms. After discovering Glucose Goddess, I feel so much better – I'm healthier, happier, and my energy is back! A huge thank-you.' — *Christie, 37*

'Over the last two years my hair was falling out like crazy. I was confused and devastated. And then a miracle happened: I followed the Glucose Goddess principles for 40 days and now it's growing back fuller and thicker! I am so happy! Not just that, but I've reversed my prediabetes too (I used to have 6 mmol/L fasting glucose levels, now they are 5.3). My energy is so much more stable during the day. I no longer need that second cup of coffee in the afternoon or that 'emergency' snack. My mental clarity has improved and my adult acne went away. It's amazing how quickly the changes happened. I recommend Jessie to everyone I know.' —*Aya, 27*

'I have type 1 diabetes. For decades no one could help me with it. Since discovering Glucose Goddess, my cravings disappeared, I was finally able to follow a healthier diet, my glucose went from 29 mmol/L to 9 mmol/L in the first days and my insulin dose divided by 10. Oh, and I lost 6lb! My doctor and nutritionist were so surprised and now they recommend Glucose Goddess to their patients.' —*Mariel, 43*

# GLUCOSE REVOLUTION

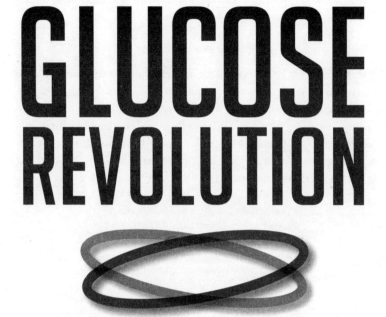

## The life-changing power
## of balancing your blood sugar

# JESSIE INCHAUSPÉ

## Jessie's disclaimer

In this book, I make existing scientific discoveries accessible to everyone. I translate them into practical tips. I am a scientist, not a doctor, so remember that none of this is medical advice. If you have a medical condition or take medication, speak to your doctor before using the hacks in this book.

## Publisher's disclaimer

The material in this book is for informational purposes only. As each individual situation is unique, you should use proper discretion, in consultation with a health care practitioner, before undertaking the diet, exercise, and techniques described in this book. The author and publisher expressly disclaim responsibility for any adverse effects that may result from the use or application of the information contained in this book.

Published in 2022 by Short Books
an imprint of Octopus Publishing Group Ltd
Carmelite House, 50 Victoria Embankment
London, EC4Y 0DZ
www.octopusbooks.co.uk

An Hachette UK Company
www.hachette.co.uk

12

A CIP catalogue record for this book is available from the British Library.

Illustrations copyright © Evie Dunne 2022
Cover design by Smith & Gilmour
Author photograph © Osvaldo Ponton

ISBN: 978-1-78072-523-9

Printed and bound in Great Britain by Clays Ltd, Elcograf S.p.A.

This FSC® label means that materials used for the product have been responsibly sourced

MIX
Paper from
responsible sources
FSC® C104740

To my family

# Contents

# Part III: How can I flatten my glucose curves?

# Dear Reader

What was the last thing you ate?

Go on, think about it for a second.

Did you like it? What did it look like? What did it smell like? What did it taste like? Where were you when you ate it? Who were you with? And why did you pick it?

Food is not only delicious, it is vital to us. Yet sometimes, without our knowing, food can also cause unintended consequences. So now for the harder questions: do you know how many grams of fat were added to your belly after eating that thing? Do you know if it will cause you to wake up with a pimple tomorrow? Do you know how much plaque it built up in your arteries or how many wrinkles it deepened on your face? Do you know if it's the reason you'll be hungry again in two hours, sleep poorly tonight, or feel sluggish tomorrow?

In short – do you know what the last thing you ate did to your body and mind?

Many of us don't. I certainly didn't before I started learning about a molecule called glucose.

For most of us, our body is a black box: we know its functions but not exactly how it works. We often decide what to have for lunch based on what we read or hear, rather than based on what our body truly needs. 'The animal tends to eat with his stomach, and the man with his brain,' wrote the philosopher Alan Watts. If only our body could speak to us, it would be a different story. We'd know exactly why we were hungry again in two hours, why we slept poorly last night, and why we felt sluggish the next day. We would make better decisions about what we ate. Our health would improve. Our lives would improve.

Well, I've got a scoop for you.

As it turns out, our bodies speak to us all the time.

We just don't know how to listen.

Everything we put in our mouths creates a reaction. What we eat affects the 30 trillion cells and 30 trillion bacteria within us. Take your pick: cravings, pimples, migraines, brain fog, mood swings, weight gain, sleepiness, infertility, polycystic ovarian syndrome (PCOS), type 2 diabetes, fatty liver disease, heart disease… are all messages from our bodies that there are problems within.

This is where I blame our environment. Our nutritional choices are influenced by billion-dollar marketing campaigns aimed at making money for the food industry – campaigns for fizzy drinks, fast food and sweets. These are usually justified under the guise of 'what matters is how much you eat – processed foods and sugar aren't inherently bad.' But science is demonstrating the opposite: processed foods and sugar *are* inherently bad for us, even if we don't eat them in caloric excess.

Even so, it's because of this misleading marketing that we believe statements such as:

'Weight loss is just about calories in and calories out.'
'You should never skip breakfast.'
'Rice cakes and fruit juice are good for you.'
'Fatty foods are bad for you.'
'You need to eat sugar to have energy.'
'Type 2 diabetes is a genetic disease that you can't do anything about.'

'If you aren't losing weight, it's because you don't have enough willpower.'

'Feeling sleepy at 3 p.m. is normal – drink some coffee.'

Our misled food choices influence our physical and mental well-being – and stop us from waking up every morning feeling amazing. It may not seem like much that we don't feel amazing every morning, but if you could… wouldn't you? I'm here to tell you there's a way you can.

Scientists have been studying how food affects us for a long time, and we now know more than we ever have on this topic. Exciting discoveries have happened in the past five years in labs around the world: they've revealed our body's reaction to food *in real time* – and have proven that although *what* we eat matters, *how* we eat it – in which order, combination, and grouping – matters too.

What the science shows is that in the black box that is our body, there is one metric that affects all systems. If we understand this one metric and make choices to optimise it, we can greatly improve our physical and mental well-being. This metric is the amount of blood sugar, or *glucose*, in our blood.

Glucose is our body's main source of energy. We get most of it from the food we eat, and it's then carried in our bloodstream to our cells. Its concentration can fluctuate greatly throughout the day, and sharp increases in concentration – I call them *glucose spikes* – affect everything from our mood, our sleep, our weight, and our skin to the health of our immune system, our risk for heart disease, and our chance of conception.

You will rarely hear glucose discussed unless you have diabetes, but glucose actually affects each and every one of us. In the last few years, the tools to monitor this molecule have become more readily available. That, in combination with the advancements in science I mentioned above, means that we have access to more data than ever before – and we can use this data to gain insight into our bodies.

This book is organised into three parts: 1) what glucose is and what we mean when we talk about glucose spikes, 2) why glucose spikes are harmful, and 3) what we can do to avoid spikes while still eating the food we love.

In Part I, I explain what glucose is, where it comes from, and why it's so important. The science is out there, but the news isn't spreading nearly fast enough. Regulating glucose is important for everyone, diabetes or no diabetes: 88 per cent of Americans are likely to have dysregulated glucose levels (even if they are not overweight according to medical guidelines), and most don't know it. When our glucose levels are dysregulated, we experience glucose spikes. During a spike, glucose floods into our body quickly, increasing its concentration in our bloodstream by more than 1.7 milli-moles per liter (mmol/L) in the span of about an hour (or less), then decreasing just as quickly. The spikes have harmful consequences.

In Part II, I describe how glucose spikes affect us in the short term – hunger, cravings, fatigue, worse menopause symptoms, migraine, poor sleep, difficulty managing type 1 diabetes and gestational diabetes, weakened immune system, worsened cognitive function – and in the long term. Dysregulated glucose levels contribute to aging and to the development of chronic conditions such as acne, eczema, psoriasis, arthritis, cataracts, Alzheimer's disease, cancer, depression, gut problems, heart disease, infertility, PCOS, insulin resistance, type 2 diabetes, and fatty liver disease.

If you were to plot your glucose level every minute of every day on a graph, the line between the points would have peaks and valleys. That graph would show your *glucose curve.* When we make lifestyle changes to avoid spikes, we flatten our glucose curves. The flatter our glucose curves, the better. With flatter glucose curves, we reduce the amount of insulin – a hormone released in response to glucose – in our body, and this is beneficial, as too much insulin is one of the main drivers of insulin resistance, type 2 diabetes, and PCOS. With flatter glucose curves, we also naturally flatten our fructose curves – fructose is found alongside glucose in sugary foods – which is also beneficial, as too much fructose increases the likelihood of obesity, heart disease, and non-alcoholic fatty liver disease.

In Part III, I'll show you how you can flatten your glucose curves with ten simple food hacks that you can easily incorporate into your life. I studied mathematics in college, then biochemistry in grad school, and this training has allowed me to analyse and distil a vast amount of nutritional science. In addition, I have run many experiments on myself wearing a device called a continuous glucose monitor, which shows me my glucose levels in real time. None of my hacks asks you to never eat dessert again, count calories, or exercise for hours and hours a day. Instead, they ask you to use what you've learned about your physiology in Parts I and II – really listening to your body – to make better decisions about *how* you eat. (And that often means putting *more* food on our plate than usual.) In this final section, I will arm you with all the information you'll need to avoid glucose spikes without wearing a monitor yourself.

Throughout this book, I draw on cutting-edge science to explain why these hacks work and tell real-life stories that show them in action. You will see data taken from my own experiments and experiments from the Glucose Goddess community, an online community I have built and grown that has (at printing time) over 200,000 members. And you'll read testimonials from members who have shed weight, curbed their cravings, improved their energy, cleared their skin, rid themselves of PCOS symptoms, reversed type 2 diabetes, done away with guilt, and gained immense self-confidence based on the insights here.

By the end of this book, you'll be able to listen to the messages coming from your body – and understand what to do next. You'll make empowered food decisions, no longer prey to marketing messages. Your health will improve, and so will your life.

I know this for a fact because it happened to me.

# HOW I GOT HERE

You know the saying 'Don't take your health for granted'? Well, I did, until an accident at 19 changed my life.

I was in Hawaii on vacation with a few friends. One afternoon we went for a hike in the jungle, and we decided that jumping off a waterfall would be a great idea (spoiler alert: it was not).

It was the first time I had ever tried anything like it. My friends had told me what to do: 'Keep your legs really straight so that your feet go into the water first.'

'Got it!' I said, and off I went.

Totally terrified, I forgot that advice as soon as I leapt off the edge of the cliff. I did *not* land feet first – I landed butt first. The pressure from the water created a shock wave up my spine, and, like dominos falling, each of my vertebrae compressed.

*Shack-shack-shack-shack-shack-shack-shack*, they went – all the way up to my second thoracic vertebra, which exploded into fourteen pieces under the pressure.

My life exploded into pieces, too. After that, I divided it into two: *before* the accident and *after* the accident.

I spent the next two weeks immobilised in a hospital bed, waiting to undergo spinal surgery. As I lay awake, I kept mentally picturing what was going to happen, unable to fully believe it: the surgeon was going to open my torso from the side, at my waist, then from the back, at the level of the broken vertebra. He was going to take out the bone fragments as well as the two surrounding discs, then fuse three vertebrae together and drill six three-inch metal rods into my spine. With an *electric drill*.

The risks associated with the procedure terrified me: lung perforation, paralysis, even death. It wasn't as though I had a choice, though. The vertebral pieces were pressing against the membrane of my spinal cord. Any shock (even tripping on a stair) could lead them to rupture the membrane, paralysing me from the waist down. I was scared. I imagined myself on the operating table, bleeding out, and the doctors giving up. I imagined my life ending like that, all because I had gotten scared in midair doing something that was supposed to be fun.

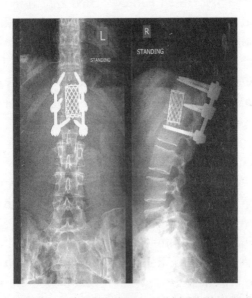

The finished result. (No, I don't set off the alarm at airport security, and yes, this stays in forever.)

The day of the surgery approached slowly but surely, though when it finally arrived, I wished it hadn't. As the anaesthetist began putting me under for the eight-hour procedure, I wondered if she would be the last person I ever saw. I prayed. I wanted to live. If I could wake up on the other side of this, I knew I would be filled with gratitude for the rest of my life.

I woke up. It was the middle of the night, and I was alone in a recovery room. At first, I felt immense relief: I was alive. Then I felt pain.

Correction: I felt a *lot* of pain. The new hardware was like an iron fist squeezing my spine. I tried to sit up to call a nurse. After a few tries he showed up, moody and dismissive. It was a dismal way to be greeted back into the world. I cried. I just wanted my mom.

It's true, I was filled with gratitude: deep, profound gratitude to be alive. But I was also in agony. My entire back was throbbing, I couldn't move an inch without feeling that my scars were going to rip open, and the nerves in my legs were on fire for days. I was allowed a shot of painkillers every three hours. Like clockwork, a nurse would come into my room, pinch the fat on my thigh, and administer the needle – alternating legs each time.

I couldn't sleep because everything hurt so much, nor eat because the opioids made me nauseous. I lost 25 pounds in two weeks. I felt at once lucky and stupid, sorry for what had happened, guilty for putting my loved ones through this, and at a loss for what to do.

My body healed in a matter of months, but then my mind and soul were the ones that needed rehab. I felt disconnected from reality. When I looked at my hands, they didn't seem like mine. When I looked in the mirror, I was terrified. Something was wrong. But I didn't know what.

Unfortunately, no one else did, either. From the outside I seemed well again. So I kept my suffering to myself. When someone asked me how I was, I answered, 'I'm great, thanks.' If I was being honest, though, I would have responded, 'I feel like a stranger in my own body, I can't look in the mirror without losing my mind, and I'm scared to death that I'm never going to be okay again.' That was later diagnosed as depersonalisation-derealisation disorder, a mental disorder where people can't connect to themselves or the reality around them.

I was living in London at the time, and I remember sitting on the Tube, looking at the commuters opposite me, wondering how many of them were also going through something difficult and hiding it, just like I was. I dreamed that someone on the train would recognise my suffering and tell me that they understood – that they had felt as I had and come back to themselves. But of course, in vain. The people sitting three feet away had no idea what was going on inside of me. *I* barely knew what was going on inside of me. And I had no idea what was going on inside of them and whether or not they were suffering, too.

It became abundantly clear to me that it's hard to know what is going on inside our bodies. Even when we can give voice to our emotions – gratitude, pain, relief, sadness, and more – we must then find out why we feel them. Where do we start when we don't feel okay?

I remember telling my best friend, 'Nothing matters – not school, not work, not money, nothing matters more than being healthy.' It was the deepest conviction I had ever felt.

And that was how, four years later, I ended up on the train headed 39 miles south of San Francisco, to an office in Mountain View. Having decided to figure out how to communicate with my body, I felt that I needed to work at the forefront of health technology. In 2015, that forefront was genetics.

I had landed an internship at the start-up 23andMe (so named because we all have 23 pairs of chromosomes that carry our genetic code). And I wanted to be there more than I had ever wanted to be anywhere.

My thinking went like this: my DNA created my body, so if I can understand my DNA, I can understand my body.

I worked as a product manager. I had two degrees under my belt and a passion for making complicated subjects simple. I was putting those to good use: I was in charge of explaining genetic research to our customers and encouraging them to participate by answering surveys. We collected data as it had never been done before: digitally, online, on millions of people at once. Each customer was a citizen scientist, contributing to advancing our collective understanding of DNA. The goal was to innovate in the field of personalised medicine and deliver health recommendations unique to each individual.

It was the best place, with the best people, the best data, and the best mission. The atmosphere at the office was electric.

I grew close to the other scientists on the research team, then read through all the papers they had published and started asking questions. But to my disappointment, little by little, it became clear to me that DNA wasn't as predictive as I had thought. For instance, your genes can increase your likelihood of developing type 2 diabetes, but they can't tell you for sure whether you'll get it. Looking at your DNA can only give you a sense of what *might* happen. For most chronic conditions, from migraines to heart disease, the cause ends up being much more attributable to 'lifestyle factors' than to genetics. In short, your genes don't determine how you feel when you wake up in the morning.

In 2018, 23andMe launched a new initiative. It was led by the Health Research & Development team, which was in charge of coming up with cutting-edge ideas. They were discussing... *continuous glucose monitors*.

Continuous glucose monitors (CGMs) are small devices worn on the back of your arm that track glucose levels. They were created to replace the finger pricks that people with diabetes have been using for decades and that give glucose measurements only a few times a day. With a CGM, glucose levels are measured every few minutes. Now entire glucose curves are revealed and conveniently sent to your smartphone. It was a real game changer for those with diabetes, who rely on glucose measurements to dose their medication.

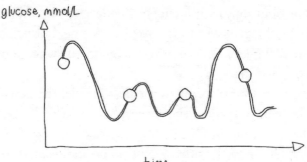

**Continuous glucose monitors, or CGMs (the line), capture glucose curves that traditional finger prick tests (white circles) miss.**

Soon after 23andMe launched the project, top athletes started to wear CGMs, too, using glucose measurements to optimise their athletic performance and endurance. And then a few scientific papers were published on studies using the devices to show that people without diabetes could have highly dysregulated glucose levels, too.

When the Health Research & Development team announced a new study looking into food response in nondiabetics, I immediately asked to be a part of it. I was always on the lookout for something that could help me understand my own body. But I definitely did not expect what came next.

A nurse came to our office to apply the device to the four of us who had volunteered. We waited for her in a glass-walled conference room; then we literally rolled up our sleeves. After wiping the back of my left upper arm with an alcohol swab, the nurse placed an applicator against my skin. I was told that a needle would go in and insert a tiny 3-millimeter-long fibre (an electrode) under my skin. Then the needle would come out, leaving the fibre in place and an adhesive transmitter on top of it. It would stay in for two weeks.

One, two… click! The monitor was in – and it was almost painless.

The sensor needed 60 minutes to start up, but then, with my phone handy, I could check my glucose levels at any time.* The numbers showed me how my body responded to what I ate (or didn't) and how I moved (or didn't). I was getting messages from the *inside*. Well, hello there, body!

When I felt great, I checked my glucose. When I felt terrible, I checked my glucose. When I worked out, when I woke up, when I went to sleep, I checked my glucose. My body was talking to me through the spikes and dips on my iPhone screen.

I ran my own experiments and took note of everything.

---

* Technically not in my blood but in the fluid between my cells. These are tightly correlated.

My lab was my kitchen, my test subject was myself, and my hypothesis was that food and movement influence glucose through a set of rules that we could define.

Quite quickly, I started noticing strange patterns: nachos on Monday, big spike. Nachos on Sunday, no spike. Beer, spike. Wine, no spike. M&Ms after lunch, no spike. M&Ms before dinner, spike. Tired in the afternoon: glucose had been high at lunch. Lots of energy all day: glucose was very steady. Big night out with friends: glucose rollercoaster through the night. Stressful presentation at work: spike. Meditation: steady. Cappuccino when I was rested: no spike. Cappuccino when I was tired: spike. Bread: spike. Bread and butter: no spike.

Things got even more interesting as I linked my mental states to my glucose levels. My brain fog (which I had started experiencing since my accident) often correlated with a big spike, sleepiness with a big dip. Cravings correlated with a glucose rollercoaster – spikes and dips in quick succession. When I woke up feeling groggy, my glucose levels had been high throughout the night.

I sifted through the data, reran many experiments, and checked my hypothesis against published studies. To feel my best, it became clear that I had to avoid big spikes and dips in my glucose levels. And that was what I did: I learned how to flatten my glucose curves.

I was making transformative discoveries about my health. I cured my brain fog and curbed my cravings. When I woke up, I felt amazing. For the first time since my accident, I began to feel truly well again.

So I started telling my friends about it. That's how the Glucose Goddess movement began.

At first, I got a lot of blank stares. I showed my friends the studies and told them that they, too, should care about flattening their glucose curves. Silence.

It became clear that I had to find a way of communicating those studies in an engaging way. I thought about using my own glucose data to illustrate the science. The problem was that at the outset, the insights from it were difficult to grasp.

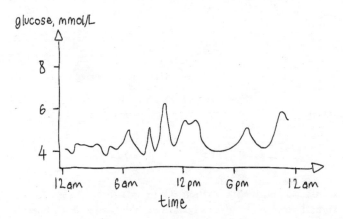

**A day's glucose data, right off the continuous glucose monitor.
It's not clear what is going on here.**

To be able to make sense of it, I needed to 'zoom in' to a specific time of day. But there was no way of doing so in the app the continuous glucose monitor came with. So I built software on my computer to do it myself.

I started keeping a diary of everything I ate. For each entry in my diary, I zoomed into four-hour windows. For instance, '5:56p.m. – glass of orange juice'. I looked at my glucose measurements starting one hour before I drank the juice and ending three hours later. Those gave me a convenient view of where my glucose levels were before I drank it, during, and after.

To make it easier on the eyes, I turned the dots into a line and filled in the spike.

And then, because science should be stylish, too, I

simplified the axis and added an image of the food on the right. That was certainly more engaging.

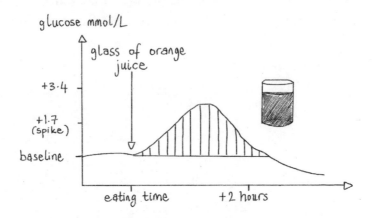

**A finished graph with my homemade software. Orange juice, and all other fruit juices, contain no fibre and a lot of sugar. Drinking them leads to a glucose spike.**

My friends and family were fascinated by the graphs. They asked me to test more and more foods and share the results. And then they started getting their own continuous glucose monitors. They sent me their data, and I aggregated it. One thing led to another, and after a while I didn't have enough time to keep up with the graph-making demand – so I built a phone app that automated it. My friends started using the app, friends of friends, too… it caught on like wildfire. Even friends without CGMs, buoyed by the evidence, started changing their eating habits.

And then, in April 2018, I started the @glucosegoddess Instagram account, and as the community grew and responded to my experiments and sent me results from their own, I became increasingly startled. Glucose, I realised, was associated with just about everything.

PART I

# WHAT IS GLUCOSE?

# Chapter 1

# Enter the cockpit: why glucose is so important

Navigating our health sometimes feels like glancing into the cockpit of a plane on the way to our seat. We see complicated stuff everywhere: screens, dials, levers, flashing lights, knobs, switches… buttons to the left, buttons to the right, buttons on the ceiling (no, but really, why do they have buttons on the *ceiling*?). We look away feeling grateful that the pilots know what they are doing. As passengers, all we care about is whether or not the plane stays in the air.

When it comes to our bodies, we are the clueless passengers, but – plot twist – we're also the pilots. And if we don't know how our bodies work, it's as if we're flying blind.

We know how we want to feel. We want to wake up with a smile, feeling energised and excited for the day. We want to have a skip in our step, feel pain free. We want to spend quality time with our loved ones, feeling positive and grateful. But it can be challenging to know how to get there. We're overwhelmed by all the buttons. What to do? Where to start?

We should start with glucose. Why? Because it's the lever in the cockpit with the biggest bang for its buck. It's the easiest to learn about (thanks to continuous glucose monitors), it affects how we feel *instantaneously* (because it influences our hunger and mood), and many things fall into place once we get it under control.

If our glucose levels are out of balance, dials flash and alarms go off. We put on weight, our hormones get out of whack, we feel tired, we crave sugar, our skin breaks out, our hearts suffer. We inch closer and closer to type 2 diabetes. If our body is the plane, the symptoms are the pitch, roll and yaw of a machine out of control. And these strongly indicate that we need to rectify something to avoid a crash. To get back into ideal cruising mode, we need to flatten our glucose curves.

How do we move this lever? Very easily – with what's on our plate.

## Yes, this book is for you

A recent study showed that only 12 per cent of Americans are metabolically healthy, which means that only 12 per cent of Americans have a perfectly functioning body – including healthy glucose levels. We don't have this exact number for all countries, but we know that metabolic health and glucose levels are getting worse everywhere. Odds are that *you*, and nine out of the ten people closest to you, are on a glucose rollercoaster without knowing it.

Here are some questions to ask yourself to find out if your glucose levels are dysregulated.

- Have you been told by a doctor that you need to lose weight?
- Are you trying to lose weight but finding it difficult?
- Is your waist size (or trouser size) above 40 inches if you are a man or above 35 inches if you are a

woman? (Waist size is better for predicting underlying disease than body mass index, or BMI, is.)

- Do you have extreme hunger pangs during the day?
- Do you feel agitated or angry when you are hungry, aka *hangry*?
- Do you need to eat every few hours?
- Do you feel shaky, lightheaded, or dizzy if meals are delayed?
- Do you crave sweet things?
- Do you feel sleepy mid-morning or mid-afternoon or are you tired all the time?
- Do you need caffeine to keep you going throughout the day?
- Do you have trouble sleeping or wake up with heart palpitations?
- Do you have energy crashes where you break out in a sweat or get nauseous?
- Do you suffer from acne, inflammation, or other skin conditions?
- Do you experience anxiety, depression, or mood disorders?
- Do you experience brain fog?
- Is your mood variable?
- Do you frequently get colds?
- Do you experience acid reflux or gastritis?
- Do you have hormonal imbalances, missed periods, PMS, infertility or PCOS?
- Have you ever been told that your glucose levels are elevated?
- Do you have insulin resistance?
- Do you have prediabetes or type 2 diabetes?
- Do you have non-alcoholic fatty liver disease?
- Do you have heart disease?
- Do you have difficulty managing gestational diabetes?

- Do you have difficulty managing type 1 diabetes?

And most important: do you think you could feel better than you currently do? If the answer is yes, keep reading.

## What this book says – and what it doesn't

Before we dive in, it's important to know which conclusions *not* to draw from this book. Let me explain.

As a teenager, I went on a vegan diet. It was a *bad* vegan diet – instead of cooking nutrient-rich chickpea stews and loading up on crispy baked tofu and steamed edamame, I chose (vegan) Oreos and (vegan) pasta. All I ate was poor-quality, glucose-spiking food. My skin broke out in pimples, and I was constantly tired.

As a young adult, I went on a keto diet. It was a *bad* keto diet. I had hoped to lose weight; instead I *gained* weight because in the process of removing all carbohydrates from my diet, all I ate was cheese. I stressed my hormonal system so much that my period stopped.

The more I've learned, the more I've realised that there is no benefit to extreme diets – especially because dogmas can easily be abused (there is very unhealthy vegan food, and there is very unhealthy keto food). The 'diets' that work are the ones that flatten our glucose, fructose and insulin curves. When vegan and keto are done well, they both do this. And when any diet is done well – meaning that it helps you reverse disease or lose excess weight – it's for that same reason. Really, we should be looking for sustainable lifestyles, not diets, and there is space on all of our plates for a little bit of everything – including sugar. Knowing how glucose works has helped me understand that better than ever.

On the topic of being moderate, I want to note three important things to keep in mind as you read this book.

First, **glucose isn't everything**. Some foods will keep your glucose levels completely steady but aren't great for your health. For instance, industrial processed oils and trans fats age, inflame and hurt our organs, but they don't cause glucose spikes. Alcohol is another example – it doesn't spike our glucose levels, but that doesn't mean it's good for us.

Aside from glucose, there are other factors that determine our health: sleep, stress, exercise, emotional connection, medical care and more. Beyond glucose, we should pay attention to fat, to fructose and to insulin, too. I'll get to these later in this book. But both fructose and insulin levels are hard to monitor continuously. Glucose levels are the only measure we can track from the comfort of the couch, and the good news is that when we flatten our glucose curves, we also flatten our fructose and insulin curves. This is because fructose exists only hand in hand with glucose in foods and because insulin is released by our pancreas in response to glucose. When the numbers on insulin are available in the scientific studies (insulin is often measured continuously in clinical settings), I describe the effect of the hacks on them, too.

Second, **context is key**. My mother often sends me a photo of something she is debating buying at the supermarket. 'Good or bad?' she texts. I always respond, 'It depends – what would you eat instead?'

We can't say whether a food is good or bad in a vacuum – everything is relative. High-fibre pasta is 'good' compared to regular pasta but 'bad' compared to veggies. An oatmeal cookie is 'bad' in relation to almonds but 'good' in relation to a can of Coca-Cola. You see the conundrum. You cannot look at a single food's glucose curve and determine whether it is 'good' or 'bad'. You must compare it to an alternative.

Finally, the **recommendations here are always based on evidence**. Every glucose graph in this book is here to illustrate scientific discoveries that I reference and cite. I do not draw generalised conclusions from a single person's glucose experiments or indeed from my personal experiments on their own. First, I do the research: I find scientific studies that explain how a certain habit flattens glucose curves – for example, a paper that finds that 10 minutes of moderate physical activity after a meal reduces the glucose spike of that meal. In these studies, the experiment has been run on a large group of people and the scientists have come to a generalised conclusion that statistically holds true. All I want to do is make a visual example of what they have found. So I pick a popular food that spikes glucose levels when eaten alone, such as a bag of crisps. Then I eat the bag of crisps on its own one morning, measure the resulting glucose curve, and do the same thing the next morning – but then go for a 10-minute walk. The second spike is smaller, just as the paper explains. That's what I show to people to illustrate that walking after any meal reduces the glucose spike of that meal. Sometimes it's not me but someone else from the Glucose Goddess community who contributes the illustrative test.

So if your body is a plane and you're both the pilot and the passenger, consider these three caveats your safety lesson. Now that you know that flattening your glucose curves is the place to start to get your body back to cruising altitude, buckle up: it's time to begin this journey by learning where glucose comes from.

# Chapter 2

# Meet Jerry: how plants create glucose

Plants don't get enough credit. To be fair, they rarely advertise their exploits. (They can't.) But if the cactus on your desk could speak, it would impress you with the tale of its ancestors: after all, they were the ones who invented the most important biological process on Earth – photosynthesis.

Millions of years ago, our planet was a barren rock of water and mud. Life consisted only of bacteria and wiggly worms in the oceans; no trees, no chirping birds and certainly no mammals or humans.

Somewhere, in one of the corners of this blue planet, perhaps where South Africa is now, a magical thing happened. After millions of years of trial and error, a tiny sprout poked through the crust of the earth, opening one leaf and along with it a new chapter of the history of life.

Quite the feat. How did that sprout do it?

It was once common to assume that plants were 'soil eaters': that they made themselves out of dirt. In the 1640s, a Flemish scientist by the name of Jan Baptist van Helmont set out to understand whether that was truly the case. He performed a five-year-long test known as the Willow Experiment, from which humanity learned two things: first, that van Helmont was very patient; second, that plants do *not* make themselves out of dirt.

Van Helmont planted a 5-pound baby willow tree in a large pot filled with 200 pounds of soil. For the next five years, he watered it and watched it grow. Then, after those five years had passed and the tree had grown, he took the tree out of the pot and weighed it again: it stood at 169 pounds – 164 pounds heavier than it had been at the beginning. But most important, the weight of the *soil* in the pot remained virtually unchanged. That meant that the 164 pounds of tree had to have come from somewhere else.

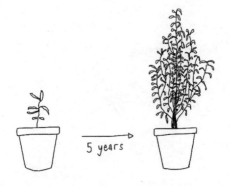

**The Willow Experiment proved that plants aren't made of dirt.**

So how do plants make their... plant stuff, if it isn't from soil? Back to the tiny sprout that just saw the light of day on Earth. Let's call him Jerry. Jerry was the first to put together a very elegant solution: the ability to transform not soil but *air* into matter. Jerry combined carbon dioxide (from the air) and water (from the soil, but not actually soil), using the energy of the sun, to make a never-before-seen substance that he used to construct every part of himself. This substance is what we now call *glucose*. Without glucose, there would be no plants and no life.

For hundreds of years after the Willow Experiment, hordes of researchers tried to understand how plants did what they did with the help of experiments involving candles, vacuum-sealed jars and many different species of algae.

The three men who finally cracked it were American scientists by the names of Melvin Calvin, Andrew Benson and James Bassham. For the discovery, Calvin was awarded the 1961 Nobel Prize in Chemistry. The process was baptised the 'Calvin-Benson-Bassham Cycle'. Since it's not the catchiest name, we commonly refer to it as *photosynthesis*: the process of turning carbon dioxide and water into glucose using the energy of the sun.

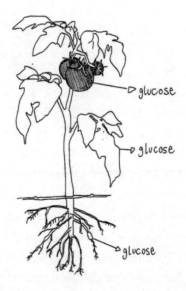

**Plants turn a sunny afternoon into glucose during photosynthesis and assemble glucose into various forms to grow. Here we see roots, leaves and fruit.**

I'm a bit envious of the way plants do what they do. They don't have to spend any time at the grocery store. They create their own food. In human terms, it would be like being able to inhale molecules from the air, sit in the sun and create a creamy lentil soup inside our stomachs without needing to find it, cook it, or swallow it.

Once created, plants can either break down glucose to use as energy or keep it intact to use as a building block. And you couldn't dream of a better brick. It's so tiny and nimble that you could fit 500,000 molecules of it into the period at the end of this sentence. It can be used to make the plant's rigid trunk, flexible leaves, spindly roots, or juicy fruit. Just as diamonds or pencil lead can be made from the exact same atom (carbon), plants can make many different things out of glucose.

## Strong starch

Among the things plants can make from glucose is *starch*. A living plant needs a supply of energy at all times. However, when it's not sunny out, either because it's cloudy or because it's dark, photosynthesis cannot take place and provide the plant with the glucose it needs to survive. In order to solve this problem, plants make extra glucose during the day and pack it away into reserves for later use.

The thing is, storing glucose isn't easy. Glucose's natural tendency is to dissolve into everything around it, like children let loose into a playground at playtime. Kids race and dash in random directions, generally uncontrollably and unpredictably, but can be rounded up by their teacher and sit (mostly) quietly behind their desks when class begins again. Similarly, plants have a solution to round up glucose. They enlist tiny helpers called *enzymes* – teacher's aides, if you will – that grab glucose molecules by the hand and attach them to each other: left hand with right hand, left hand with right hand, hundreds and thousands of times over. The result is a long chain of glucose, no longer racing and dashing in random directions.

This form of glucose is called *starch*. It can be stored in small amounts throughout the plant, but mostly in its roots.

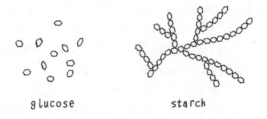

**Plants assemble glucose into long chains called *starch* in order to store it.**

Beets, potatoes, carrots, celeriac, parsnips, turnips and yams are all roots, and all contain starch. Seeds also contain starch, which provides the necessary energy to help them grow into a plant. Rice, oatmeal, corn, wheat, barley, beans, peas, lentils, soybeans and chickpeas are all seeds, and all of them contain starch, too.

**Root vegetables and seeds are packed with starch.**

Starch is strong – the name comes from the Germanic word 'stiff' or 'rigid' – but that doesn't mean it's inflexible. It can be taken apart with the right tool. Whenever plants need glucose, they use an enzyme called alpha-amylase that heads to the roots and frees some glucose molecules from their starch chains. *Snap* – glucose is let loose, ready to be used as energy or as a building block.

## Fierce fibre

Another enzyme (there are a number of them) can be called on to perform a different task. Instead of attaching glucose molecules hand to hand to make starch, this enzyme connects glucose molecules hand to foot, and the resulting chain is called *fibre*. This substance is as important as grout between the bricks of a house. It's what allows plants to grow tall without falling over. It's most commonly found in trunks, branches, flowers and leaves, but there is fibre in roots and fruit as well.

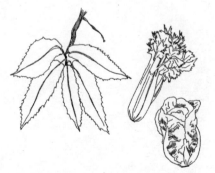

**Trunks, branches and leaves contain the most fibre.**

Humans found a practical purpose for fibre: it was

harvested and processed to create paper, from Egyptian papyruses onward. Today, it's extracted from tree trunks, polymerised and turned into sheets and reams of paper. If you're reading these words in a physical book, you're reading a book about glucose printed on glucose.

## Flirty fruit

If you were to lick glucose, it would taste sweet. But plants also transform some of their glucose into molecules that are 2.3 times as sweet, called *fructose*.

Plants concentrate fructose into fruit – apples, cherries, kiwis and more – that they dangle from their branches. The purpose of fructose is to make fruit taste irresistible to animals. Why do plants want their fruit to be irresistible? Because they hide their seeds in them. It's key to propagation: plants hope that animals will eat their fruit and their seeds will go unnoticed until they come out their eater's other end. That's how seeds spread far and wide, thereby ensuring the plants' survival.

**Fruits are full of fructose.**

Most of plants' fructose is used in this way, but some, with the help of another enzyme, links up, for a time, with glucose. The result is a molecule called *sucrose*. Sucrose exists to help plants compress energy even further (a sucrose molecule is slightly smaller than a glucose and fructose molecule side by side, which allows plants to store more energy in a tighter space). For plants, sucrose is an ingenious temporary storage solution, but for us, it has a huge significance. We use it every day, under a different name: table sugar.

Starch, fibre, fructose and sucrose – the various forms glucose can take – exist thanks to photosynthesis. And this, Jerry's elegant solution, paved the way for the rest of life on this planet.

# Chapter 3

# A family affair: how glucose gets into the bloodstream

The glucose-burning system that plants invented became vital to all living things, from dinosaurs to dolphins to mice. Four hundred and forty-nine million years after the first plant appeared, humans arrived – and they burned glucose, too.

Your cells, like all animal and plant cells, need energy to stay alive – and glucose is their prioritised energy source. Each of our cells uses glucose for energy according to its specific function. Your heart cells use it to contract, your brain cells to fire neurons, your ear cells to hear, your eye cells to see, your stomach cells to digest, your skin cells to repair cuts, your red blood cells to bring oxygen to your feet so you can dance all night long.

*Every second*, your body burns eight billion billion molecules of glucose. To put that into perspective, if each glucose molecule were a grain of sand, you'd burn every single grain of sand on all the beaches of the earth *every ten minutes*.

Suffice it to say, humans need a tremendous amount of fuel.

There is just one tiny hiccup: humans aren't plants. Even with the very best of intentions, we can't make glucose from the air and the sun (I tried to photosynthesise at the beach once – to no avail).

The most common way (but not the only way) for us to get the glucose we need is by eating it.

## Starch

When I was 11 years old, we conducted an experiment in biology class that I remember to this day. We sat down for second period, and each student was handed a slice of white bread.

As we looked around perplexed, our teacher broke the news: we were to put the whole slice in our mouth and chew it – fighting the urge to swallow – for a full minute. It was an odd request but arguably more fun than our usual class activities, so off we went.

About 30 chews in, something surprising happened: the taste of the bread started shifting – it began to taste sweet!

Starch was turning into glucose in our mouths.

A slice of bread is made, for the most part, from flour. Flour is made by grinding wheat kernels and wheat kernels, as you know, are filled with starch. Any food made from flour contains starch. Pie crust, cookies, pastries, pasta – all are composed of flour, so all are composed of starch. When we eat, we break starch down into glucose, using the same enzyme that plants use to do this task: alpha-amylase.

Starch is turned into glucose extremely quickly in our body. In general, the process happens mostly in our gut, where it goes unnoticed. The alpha-amylase enzymes snap the bonds of the chain, and glucose molecules are freed. There they are, running around in the playground once again.

The enzymes that do this vital work also exist in our saliva. When we chew starch long enough, we give the enzymes the time they need to begin their work. That process begins in our mouth, and we can taste it. Hence the power of this experiment.

## Fruit

Fruit, in contrast, tastes sweet from the get-go. This is because it already contains unchained glucose molecules, which taste sweet, as well as fructose, which tastes even sweeter, and their combined form, sucrose, which is sweeter than glucose but not as sweet as fructose.

Glucose from fruit is ready to be used and does not need to be snapped. Sucrose *does* need to be snapped, and there's an enzyme that separates it into glucose and fructose molecules, and this does not take long – it happens in a nanosecond.

Fructose is a little more complicated. After we eat it, a portion of it is turned back into glucose in our small intestine. The remainder of it stays in fructose form. Both permeate the lining of our gut to enter our bloodstream. I'll explain what happens next later in this book, but what I want you to remember now is that although glucose is needed to fuel your body's systems, fructose isn't. We eat a lot of unnecessary fructose in our diet nowadays, because we eat a lot more sucrose (which, as a reminder, is half glucose, half fructose).

What about fibre? Well, it has a special fate.

## Fibre

Enzymes work to snap the bonds of starch and sucrose, but there is no enzyme that can snap the bonds of fibre. It doesn't get turned back into glucose. This is why when we eat fibre, it remains fibre. It travels from our stomach to our small and large intestines. And this is a good thing. Though it doesn't turn back into glucose and therefore can't provide energy to our cells, fibre is an essential part of our diet and plays a very important role in aiding digestion, maintaining healthy bowel movements, keeping our microbiome healthy and more.

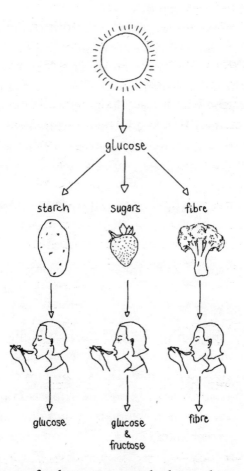

**Any part of a plant we eat turns back into glucose (and fructose) as we digest it, except for fibre, which passes right through us.**

# One parent, four siblings

Starch, fibre, fructose and sucrose are like four siblings with different personalities. They're all related, because they have the same parent, glucose – no matter how much they argue about who borrowed whose clothes.

It would almost make sense to give them a family name.

In 1969, a cohort of scientists wrote a 20-page-long document titled 'Tentative Rules for Carbohydrate Nomenclature, Part I, 1969' and presented it to the scientific community.

After that paper, it was accepted that the name for this family would be 'carbohydrates'. Why carbohydrates? Because it refers to things that were made by joining carbon (carbo) and water (hydrate), which is what happens during photosynthesis.

You may have heard of this family under its popular nickname, *carbs*.

Carbohydrates = Starch + Fibre + Sugars (glucose, fructose, and sucrose)

You'll notice that within the carbohydrate family, scientists decided to make a subgroup for the smallest molecules: glucose, fructose and sucrose. This subgroup is called *sugars*. The scientific word *sugars* is not the same as our common table sugar, even though the *sugars* group does include the molecule that constitutes table sugar, sucrose. That's scientific nomenclature for you.

Members of the carbohydrate family exist in various proportions in a plant. For example, broccoli contains a lot of fibre and some starch, potatoes contain a lot of starch and some fibre, and peaches contain mostly sugars and some fibre (you'll notice that there is at least *some* fibre in every plant).

But a bit confusingly, when people talk about nutrition, they often say 'carbohydrates' or 'carbs' to describe only starch and sugars. They don't include fibre, because it isn't absorbed into our bloodstream as its siblings are. You might hear something such as 'Broccoli has few carbs but a lot of fibre.' According to the scientific nomenclature, the correct thing to say would be 'Broccoli contains a lot of carbs, most of which are fibre.'

I will stick to convention here, because it's most likely what you'll hear from those around you. (But as always, I wanted you to understand the science!) When I say 'carbohydrates' or 'carbs', I'll be talking about starchy foods (potatoes, pasta, rice, bread and so on) and sugars (fruit, pies, cakes and more) but not about vegetables, because they contain mostly fibre and very little starch. And I will say 'sugar' when I'm referring to table sugar, as most of us do.

## What if there were no glucose in our diet?

Since glucose is so important to life, you may wonder how some carnivorous animals survive. After all, many animals don't eat plants (for instance dolphins, which feast on fish, squid and jellyfish), and some humans evolved in areas without a fruit or vegetable in sight, as in the freezing Russian plains, so they didn't eat plants, either.

Well, because glucose is so important to our cells, if we can't find any to eat, our body can *make it from within*. That's right, we don't photosynthesise and make glucose out of air, water and sunlight, but we can make glucose from the food we eat – from fat or protein. Our liver, through a process called *gluconeogenesis*, performs this process.

What's more, our bodies adapt even further: when glucose is limited, many cells in our body can, when needed, switch to using fat for fuel instead. This is called *metabolic flexibility*. (The only cells that always rely on glucose are red blood cells.)

Indeed, some diets such as Atkins and keto deliberately restrict the consumption of carbohydrates in order to keep a person's glucose levels extremely low and thus push the body into burning fat for fuel. This is called *nutritional ketosis* and is metabolic flexibility in action.

So, sure, carbohydrates aren't biologically *necessary* (we don't need to eat sugar to live), but they are a quick source of energy and a delicious part of our diet, and they have been consumed for millions of years. Scientists know that humans' prehistoric diet included both animals and plants: when plants were available, humans consumed them. What they ate depended on where they lived. They adapted to the unique food supply around them. And our food supply today looks quite different from what nature had planned.

# Chapter 4

# Seeking pleasure: why we eat more glucose than before

Nature intended us to consume glucose in a specific way: in plants. Wherever there was starch or sugar, there was fibre as well. This is important, *because the fibre helped to slow our body's absorption of glucose.* You'll learn how to use this information to your advantage in Part III.

Today, however, the vast majority of supermarket shelves are packed with products that contain mostly starch and sugar. From white bread to ice cream, sweets, fruit juices and sweetened yoghurts, fibre is nowhere to be seen. And this is on purpose: fibre is often removed in the creation of processed foods, because its presence is problematic if you're trying to preserve things for a long time.

Let me explain – and I have to admit, strawberries were harmed in the making of this example. Place a fresh strawberry in the freezer overnight. The next morning, take it out to thaw on a plate. If you try to eat it, it will be mushy. Why? Because the fibre was broken into smaller bits by the freezing and thawing process. The fibre is still there (and still has health benefits), but the texture is not the same.

**A fresh strawberry; then what it looks like after it is frozen overnight and thawed.**

Fibre is often removed from processed foods so that they can be frozen, thawed and stored on shelves for years without losing their texture. Take, for example, white flour: fibre is found in the germ and bran (outer husk) of the wheat kernel that is stripped away during milling.

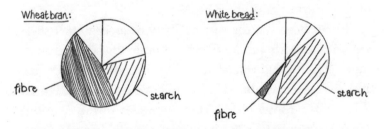

**When the starchy parts of plants are processed to make supermarket goods, they are stripped of their fibre. Fibre-packed seeds and roots are turned into starchy bread or crisps (and sugar is usually added).**

Something else is done to foods to turn them into successful supermarket products: their sweetness is increased. The basis of food processing is to first strip away the fibre, then concentrate the starch and sugars.

When we humans find something good, we tend to take it to the extreme. The smell of fresh roses pleases our senses, so thousands of tons of rose petals are distilled and concentrated into essential oil, bottled and made available anywhere, anytime, by the perfumery industry. Similarly, the food industry wanted to distil and concentrate nature's most sought-after taste: sweetness.

You may wonder: why do we like sweetness so much? It's because in Stone Age times the taste of sweetness signalled foods that were both safe (there are no foods that are both sweet and poisonous) *and* packed with energy. In a time when food wasn't easy to find, it was an advantage to eat all the fruit before anyone else could, so we evolved to feel pleasure when we tasted something sweet.

When we do, a hit of a chemical called dopamine floods our brain. This is the same chemical that is released when we have sex, play video games, scroll social media or, with more dangerous consequences, drink alcohol, smoke cigarettes, or use illegal drugs. And we can never get enough of it.

In a 2016 study, mice were given a lever with which they could activate their own dopamine neurons (thanks to a special optical sensor). The researchers saw a peculiar behaviour: if they left the mice to their own devices, the mice spent all their time pressing the lever to activate their dopamine neurons, over and over. They stopped eating and drinking – to the point where eventually the researchers had to end the experiment because otherwise, the mice were going to die. The mice's dopamine obsession had made them forget their basic needs. This is all to say that animals, including humans, *really* like dopamine. And eating sweet foods is an easy way to get a hit.

Plants have been concentrating glucose, fructose and sucrose in their fruit forever, but a few millennia ago, humans began to do the same: we started breeding plants so that, among other reasons, their fruit would taste even sweeter.

Ancestral bananas (top image) are as nature intended them to be: full of fibre, with a small amount of sugar. The 21st-century banana (bottom image) is the result of many generations of breeding to reduce fibre and increase sugar.

On the left, a peach as it was 6,000 years ago. On the right, a 21st-century peach. The fruits we eat today are bigger and sweeter than they were thousands of years ago.

And then, by boiling sugarcane and crystallising its juice, humans created table sugar – 100 per cent sucrose. This new product became very popular in the 18th century. As the demand grew, so did the horrors of slavery: millions of slaves were taken to humid parts of the world to farm sugarcane and produce table sugar.

Sugar sources changed over time – we now extract sucrose from beets and corn, too – but regardless of which plant is used, the resulting sucrose added to processed foods is the chemical copy of the one found in fruit. What's different is its concentration.

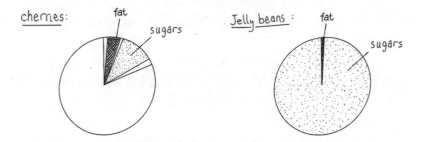

**Fruit, such as cherries, and sweets, such as jelly beans, both contain sugar. But jelly beans contain a superconcentrated amount of it.**

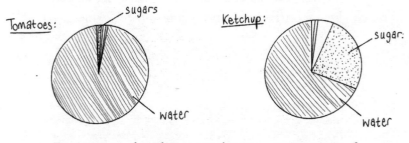

**Even tomatoes have been turned into a sweeter version of themselves: ketchup.**

Sugar has become ever more concentrated and available: we have gone from eating in-season foraged fibrous fruit in prehistoric times to eating minuscule quantities of sucrose in the 1800s (you'd have been lucky if you came across a single chocolate bar in your entire life) to eating more than *94 pounds* of the stuff per year today.

We keep eating more of it because it's hard for our brain to curb its cravings for things that taste like fruit. Sweetness and dopamine feel forever rewarding.

As the mice experiment shows, it's important to understand that the inclination to reach for a chocolate isn't our fault. It's not a willpower issue – far from it. Deep, old evolutionary programming tells us that eating Skittles is a good move.

Sheryl Crow sings that if something makes you happy, 'it can't be that bad'. We need glucose to live, and it gives us pleasure. So it's fair to wonder: what's the big deal if we eat more?

In some cases, more isn't necessarily better. Give a plant too much water, and it will drown; give humans too much oxygen, and they pass out. Similarly, there is an amount of glucose that is *just right* for us: just enough for us to feel great, jump around, go to work, hang out with other humans, live, laugh and love. But we can have too much glucose. And too much glucose hurts us, often without our realising it.

# Chapter 5

# Underneath our skin: discovering glucose spikes

A long time ago, way before I knew about glucose, I ate a Nutella crepe every morning before school. I would wake up 20 minutes before I had to leave the house, pull on jeans and a T-shirt, forget to brush my hair (sorry, Mom), head to the kitchen, grab crepe mix from the fridge, place a slab of butter on a hot pan, pour the mix onto it, *swish swish*, flip it, plate it, lather it with Nutella, fold it and eat it.

I would say goodbye to my mom, who was enjoying her own breakfast: a bowl of Special K and milk sprinkled with table sugar, and a glass of orange juice.

Millions of people were eating a similar breakfast. On the table was a display of some pretty cool technology. For me: wheat milled and turned into flour; sucrose, hazelnuts, palm oil and cocoa blended into a spread. For my mom: grains of corn popped and turned into flakes; beets squished and puréed and dried into sucrose; and oranges juiced into a liquid consisting mostly of glucose and fructose.

All that concentrated sugar tasted very sweet. Our tongues whole-tonguedly approved of the party.

The starch and sugar turned into glucose after we swallowed them; they landed in our stomach, then entered our small intestine. There, the glucose disappeared through the lining of our gut and moved into our bloodstream. From our capillaries – tiny blood vessels – it moved to larger and larger vessels, just like taking the sliproad to the motorway.

When doctors measure the amount of glucose in our body, they often draw our blood and assess its concentration there. But glucose doesn't just stay in our blood. It seeps into every part of us, and it can be measured anywhere.

That's why with a continuous glucose monitor (CGM), I can measure the amount of glucose throughout my body without a blood test: the CGM senses the concentration of glucose between the fat cells on the back of my arm.

To quantify the concentration of glucose, we use millimoles per litre, also written mmol/L. Other countries use milligrams per decilitre (mg/dL). Whatever unit is used, they refer to the same thing: how much glucose is freely roaming in the body.

The NHS states that a baseline concentration (also known as your *fasting level*, that is, your glucose level first thing in the morning before eating) between 4.0 and 5.4 mmol/L is 'normal'; that between 5.5 and 6.9 mmol/L indicates prediabetes; and anything above 7.0 mmol/L indicates diabetes.

But what the NHS describes as 'normal' may not actually be optimal. Early studies showed that the *thriving* range for fasting glucose may be between 4 and 4.7 mmol/L. That's because there is more likelihood of developing health problems from 4.7 mmol/L and up.

Furthermore, though our fasting level gives us information on whether we are at risk for being diagnosed with diabetes or not, it isn't the only thing to consider. Even if our fasting level is 'optimal', we may still experience glucose spikes on a daily basis. As explained earlier, spikes are rapid increases and drops in glucose concentration after we eat, and they are harmful. I'll explain why in the next chapter.

The NHS states that our glucose levels shouldn't increase

above 7.9 mmol/L after eating. But again, that's 'normal', not optimal. Studies in people without diabetes give more precise information: we should strive to avoid increasing our glucose levels by more than 1.7 mmol/L after eating. So in this book I will define a glucose spike as an increase in glucose in our body of more than 1.7 mmol/L after eating.

The goal is to avoid spikes, whatever your fasting level is, because it's the *variability* caused by spikes that is most problematic. It's *years* of repeated daily spikes that slowly increase our fasting glucose level, a pattern we discover only once that level is classified as prediabetic. By then, the damage has already started.

Every morning, my mom's breakfast led to a massive glucose spike of 4.6 mmol/L, taking her fasting level of 5.4 mmol/L all the way up to 10 mmol/L! That increase was well above our 1.7 mmol/L definition of a spike and even well above the 7.9 mmol/L NHS cut-off for a 'normal' spike after a meal.

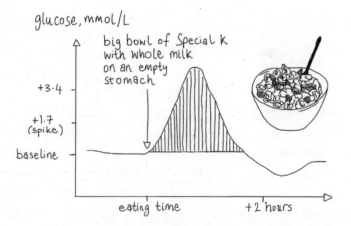

**The traditional cereal breakfast, thought of as healthy, makes our glucose spike way beyond the healthy range and then come crashing down just as quickly.**

Remember that the measurements of the concentration of glucose in your body over time, when plotted, create a glucose curve. For example, if I look at my glucose levels for the past week, my curve will be variable if I experienced many spikes or flat if I experienced fewer of them.

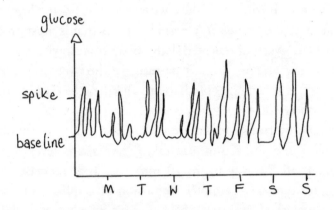

**Above, a week's glucose curves with many spikes; below, a week with fewer spikes.**

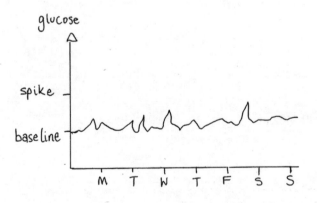

In this book, I advise you to flatten your glucose curves, which means zooming out and seeing fewer and smaller spikes over time. Another way to describe flattening your glucose curves is *reducing glycaemic variability*. The smaller your glycaemic variability, the better your health will be.

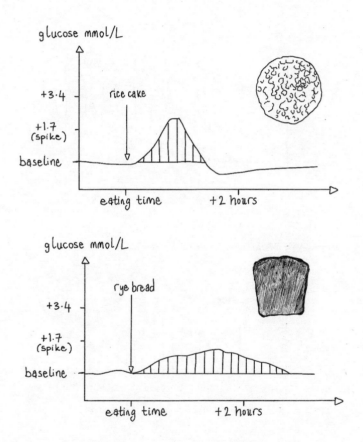

**When comparing two curves, you don't need to do any maths. The one with the taller spike, i.e. the most variability (top graph), is worse for your health.**

## Some spikes are worse than others

The two glucose spikes charted below look exactly the same. But one was more harmful than the other. Can you guess which one?

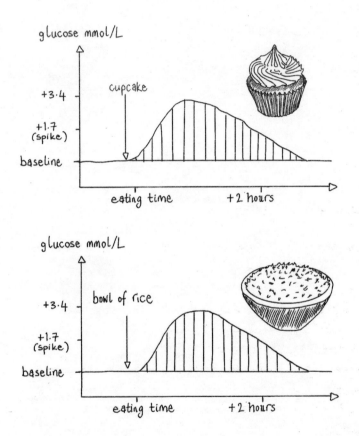

A glucose spike from a sweet food (cupcake) is worse for our health than a glucose spike from a starchy food (rice). The reason has nothing to do with the glucose measured, though; it has to do with a molecule that's not visible.

A sweet food contains table sugar, or sucrose – that

compound made up of glucose and fructose. A starchy food doesn't. Whenever we see a glucose spike from a sweet food, there is a corresponding fructose spike that unfortunately we can't see. Continuous glucose monitors can detect only glucose, not fructose, and continuous fructose monitors don't exist yet.

Until they do, remember that if the food you ate was sweet and it created a glucose spike, it also created an invisible fructose spike, and that's what makes a sweet spike more harmful than a starchy spike.

Now it's time to get to the *why*: why exactly are glucose spikes bad for us, and why are fructose spikes worse? What do they do inside our body? Put your glasses on, grab a beverage and get comfortable. By the end of Part II you'll have learned your body's language.

Part II

# WHY ARE GLUCOSE SPIKES HARMFUL?

# Chapter 6

# Trains, toast and Tetris: the three things that happen in our body when we spike

Each of us is made up of more than 30 trillion cells. When we spike, they all feel it.

Glucose's primary biological purpose once it enters a cell is to be turned into energy. The powerhouses responsible for this are microscopic organelles found in most of our cells called *mitochondria*. Using glucose (and oxygen from the air we breathe), they create the chemical version of electricity to give each cell the power to do whatever it needs to. As glucose floods into our cells, it heads straight to the mitochondria to undergo its transformation.

## Why the train stops: free radicals and oxidative stress

To understand how mitochondria respond to a glucose spike coming their way, imagine this: your grandfather, finally retiring after a long career, is able to fulfil his dream of working on a steam train. Everyone in the family thinks he's crazy for doing it, but he doesn't care. After some training, he enlists as a stoker in a train's engine room: his job is to shovel coal onto the fire to generate the steam that pushes the pistons and makes the wheels of the train turn. He's the mitochondria of the train, if you will.

Periodically throughout the day, as the train speeds along the tracks, coal is delivered to your grandfather. He places it by the furnace and shovels it into the flames at a steady pace to fuel the process that moves the train. The raw material is converted to energy. And when the stock is used up, another batch is promptly supplied.

Just like the train, our cells hum along smoothly when the amount of energy that is provided is equal to the amount of energy needed to function.

Now it's the second day of your grandfather's new job. A few minutes after the first delivery of coal, he gets a surprising knock on the door. More coal. He thinks, *Well, it's a bit early, but that way I'll have some extra*. He sets it aside next to the furnace. A few minutes later, another knock. More coal. And another. The knocks keep coming, and the coal keeps being delivered. 'I don't need all of this!' he says. But he's told that it's his job to burn it, and he's given no other explanation.

All day long, delivery after delivery, unnecessary coal is stuffed into his cabin. The coal being delivered far exceeds what is needed. Your grandfather can't burn the coal more quickly, so piles build up around him.

In a short while, there's coal everywhere, stacked to the ceiling. He can barely move. He can't shovel any more coal onto the fire because there is so much in the way. The train stops, and people get angry. At the end of the day, he quits, his dream sabotaged.

Mitochondria feel the same way when we give them more glucose than they need. They can burn only as much glucose as the cell needs for energy, not more. When we spike, we deliver glucose to our cells *too quickly*. The speed – or *velocity*

– at which it is delivered is the issue. Too much at once, and problems pile up.

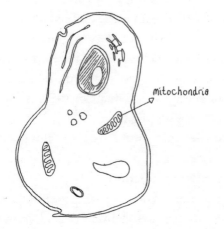

mitochondria

**A healthy cell contains thousands of functioning mitochondria, among many other components.**

According to the latest scientific theory, the Allostatic Load Model, when our mitochondria are drowning in unnecessary glucose, tiny molecules with large consequences are released by our cells: *free radicals*. (And some glucose is converted to fat; more on that shortly.) When free radicals appear because of a spike, they set off a dangerous chain reaction.

Free radicals are a big deal because anything they touch, they damage. They randomly snap and modify our genetic code (our DNA), creating mutations that activate harmful genes and can lead to the development of cancer. They poke holes in the membranes of our cells, turning a normally functioning cell into a malfunctioning one.

Under normal circumstances, we live with a moderate amount of free radicals in our cells, and we can handle them – but with repeated spikes, the quantity produced becomes unmanageable. When there are too many free radicals to be neutralised, our body is said to be in a state of *oxidative stress*.

Oxidative stress is a driver of heart disease, type 2 diabetes, cognitive decline and general ageing. And fructose increases oxidative stress even more than glucose alone. That's one of the reasons that sweet foods (which contain fructose) are worse than starchy foods (which don't). Too much fat can also increase oxidative stress.

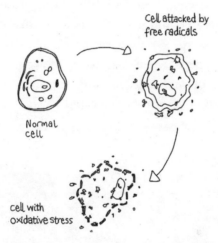

Over decades cells become ravaged. Because they're stuffed, crowded and overwhelmed, our mitochondria can't convert glucose to energy efficiently. The cells starve, which leads to organ dysfunction. We feel this as humans: even though we're fuelling up by eating, we suffer from lassitude; it's hard to get up in the morning, and we have no energy throughout the day. We're *tired*. Do you know the feeling? I sure did.

That feeling is compounded by a second process that's set off when we experience a glucose spike.

## Why you are toasting: glycation and inflammation

This may come as a surprise to you, but you are currently *cooking*. More specifically, you are *browning*, just like a slice of bread in the toaster.

Inside our body, from the moment we are born, things literally brown, albeit very slowly. When scientists look at the rib cage cartilage of babies, it's white. Once a human reaches 90 years old, that same cartilage is brown.

In 1912, a French chemist by the name of Louis-Camille Maillard described and gave his name to this phenomenon, now known as the Maillard reaction. He discovered that browning happens when a glucose molecule bumps into another type of molecule, causing a reaction. The second molecule is then said to be 'glycated'. When a molecule is glycated, it's damaged.

**When we toast bread, we make it brown. Our insides are browning just like this.**

This process is a normal and inevitable part of life, and it's why we age, why our organs slowly deteriorate and why we eventually die. We can't stop this process, but we can slow it down or speed it up.

The more glucose we deliver to our body, the more often glycation happens. Once a molecule is glycated, it's damaged forever – which is why you can't untoast a piece of toast. The long-term consequences of glycated molecules range from wrinkles and cataracts to heart disease and Alzheimer's disease. Since browning is ageing and ageing is browning, slowing down the browning reaction in your body leads to a longer life.

Fructose molecules glycate things *10 times as fast as* glucose, generating that much more damage. Again, this is another reason why spikes from sugary foods such as cookies (which contain fructose) make us age faster than do spikes from starchy foods such as pasta (which doesn't).

Glucose levels and glycation are so connected that a very well-known test to measure the level of glucose in our body actually measures glycation. The haemoglobin A1c (HbA1c) test (well known among people with diabetes) measures how many red blood cell proteins have been glycated by glucose over the past two to three months. The higher your HbA1c level, the more often the Maillard reaction is happening inside your body, the more glucose is circulating, and the faster you are ageing.

The combination of too many free radicals, oxidative stress and glycation leads to a generalised state of *inflammation* in the body. Inflammation is a protective measure; it's the result of the body trying to defend against invaders. But chronic inflammation is harmful because it turns against our own body. From the outside, you might see redness and swelling, and on the inside, tissues and organs are slowly getting damaged.

Inflammation can also be driven up by alcohol, smoking, stress, leaky gut syndrome, and substances released by body fat. Chronic inflammation is the source of most chronic illnesses, such as strokes, chronic respiratory diseases, heart disorders, liver disease and diabetes, as well as of problems like obesity. The World Health Organization calls inflammation-based diseases 'the greatest threat to human health'. Worldwide, *three out of five people will die of an inflammation-based disease*. The good news is, a diet that reduces glucose spikes decreases inflammation and along with it your risk of contracting any of these inflammation-based diseases.

The third and final process we are going to dive into might be the most surprising. It's actually a defence mechanism that our body uses to defend against spikes – but it has its own consequences.

## Playing Tetris to survive: insulin and fat gain

It's essential to our survival to get excess glucose out of circulation as quickly as possible, to reduce free radical formation and glycation. So our body, working without our even knowing it, has a plan: it starts playing a kind of Tetris.

In Tetris, players arrange blocks into rows to clear them before they accumulate. It's eerily similar to what happens in our body: as too much glucose enters, our body does its best to stash it away. Here's how it works.

When our glucose levels increase, our *pancreas* becomes the orchestra conductor of Tetris.

One of the pancreas's main functions is to send a hormone called *insulin* into the body. Insulin's sole purpose is to stash excess glucose in storage units throughout the body, to keep it out of circulation and protect us from damage. Without insulin, we would die; people without the ability to make it – those with type 1 diabetes – must inject insulin to make up for what the pancreas can't produce.

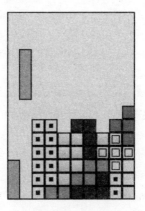

**Tetris? No – the clearing of a glucose spike.**

Insulin stashes excess glucose in several storage units. Enter storage unit number one: the *liver*. The liver is a very convenient storage unit, because all of the blood that comes from the gut carrying new glucose from digestion has to go through the liver.

Our liver turns glucose into a new form, called glycogen. It's equivalent to how plants turn glucose into starch. Glycogen is actually the cousin of starch – it's composed of many glucose molecules attached hand to hand. If excess glucose stayed in its original form, it would cause oxidative stress and glycation. Once transformed, it does no damage.

The liver can hold about 100 grams of glucose in glycogen form (the amount of glucose in two large McDonald's fries). That's half of the 200 grams of glucose that our body needs for energy per day.

The second storage unit is our *muscles*. Our muscles are

effective storage units because we have so many of them. The muscles of a typical adult weighing around 68 kilos (10st10lb) can hold about 400 grams of glucose as glycogen, or the amount of glucose in seven large McDonald's fries.

The liver and muscles are efficient, but we tend to eat much more glucose than we need, so those storage units get full rather quickly. Fairly soon, if we didn't have another storage unit for extra glucose, our body would lose its game of Tetris.

Which part of our body can we grow very easily, without much effort and just by sitting on our couch? Introducing our fat reserves.

Once insulin has stored all the glucose it can in our liver and muscles, any excess glucose is turned into fat and stored in our fat reserves. And that's one of the ways we gain weight. And then some. Because it's not just glucose that our body has to deal with, it must also dispose of fructose. And unfortunately, fructose cannot be turned into glycogen and stored in the liver and the muscles. *The only thing that fructose can be stored as is fat.*

The fat our body creates from fructose has a few unfortunate destinies: first, it accumulates in the liver and drives the development of nonalcoholic fatty liver disease. Second, it fills up fat cells in our hips, thighs and face and between our organs, and we gain weight. Finally, it enters the bloodstream and contributes to an increased risk of heart disease. (You may have heard of it as low-density lipoprotein (LDL) or 'bad' cholesterol.)

This is another reason why, if two foods have the same number of calories, I'd recommend that you skip the sweet food (which contains fructose) in favour of a savoury food (which doesn't). The absence of fructose means that fewer molecules end up as fat.

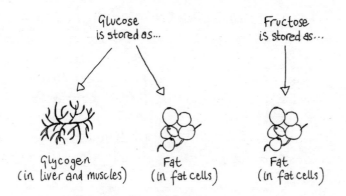

**Humans store extra glucose as glycogen and fat. Extra fructose just turns into fat.**

Ironically, many processed foods that are 'fat-free' contain a lot of sucrose, so the fructose in it is turned into fat after we digest it. More on this in Part III.

Many of us have complicated feelings about fat, but it's actually very useful: your body uses its fat reserves to provide storage space for the excess glucose and fructose floating around in your bloodstream. We shouldn't be mad at our body for putting on fat; instead, we should thank it for trying to protect us from oxidative stress, glycation and inflammation.

The more you're able to grow the number and size of your fat cells (which is usually a function of genetics), the longer you'll be protected against excess glucose and fructose (but the more weight you will put on).

Which brings me back to insulin. Insulin, as I've explained, is vital to this process, as it helps stash excess glucose in those three 'storage lockers'. And in the short term it's helpful.

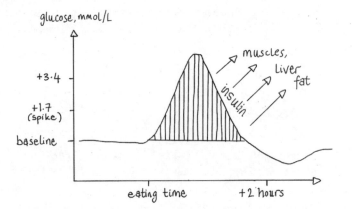

**About 60 minutes after a meal, our glucose concentration reaches its maximum and then starts coming down as insulin arrives and ushers the glucose molecules away into our liver, muscles and fat cells.**

However, the more glucose spikes we experience, the more insulin is released in our bodies. In the long term, chronically elevated levels of insulin bring problems of their own. Too much insulin is the root cause of obesity, type 2 diabetes, polycystic ovary syndrome (PCOS) and more. One of the most important things that happens when we flatten our glucose curves is that we automatically flatten our insulin curves as well.

Back to those complicated feelings about fat. It's useful, but if you're trying to shed pounds, it's important to understand what's happening in your body on a cellular level and how insulin makes things tricky. When we say 'I want to lose weight,' what we're actually saying is, 'I want to empty my fat cells of the fat they contain so that they deflate like balloons and reduce in size, bringing down my waist size.' To do so, we need to be in 'fat-burning' mode.

Just as Jerry could tap into his starch reserves at night, our body can call on glycogen in our liver and muscles to turn back into glucose whenever the thousands of mitochondria in each cell need it. Then, when our glycogen reserves begin to diminish, our body draws on the fat in our fat reserves for energy – we're in fat-burning mode – and we lose weight.

But this happens only when our insulin levels are low. If there is insulin present, our body is prevented from burning fat: insulin makes the route to our fat cells a one-way street: things can go in, but nothing can come out. We're not able to burn any existing reserves until our insulin levels start coming back down about two hours after the spike.

But if our glucose levels, and therefore our insulin levels, are steady, we shed pounds. In a 2021 study of 5,600 people, Canadian scientists showed that weight loss is always preceded by insulin decrease.

Excess glucose in our body and the spikes and dips it causes change us on a cellular level. Weight gain is just one of the symptoms we can see; there are many more. But for each, flattening our glucose curves can bring relief.

# Chapter 7

# From head to toe: how spikes make us sick

Early on, I had a profound realisation that kick-started my research into glucose – that *how I feel right now* is intimately linked to the spikes and dips of my glucose curve.

One day at work, around 11 a.m., I had grown so sleepy that I could barely move my fingers to click my mouse. Focusing on the task at hand was impossible. So, with a lot of effort, I stood up, walked to the office kitchen and poured myself a large black coffee. I drank the whole cup – and still I was exhausted. I checked my glucose levels: they had been on a steep downward slope since a big spike after a breakfast made of a salt-and-chocolate-chip cookie and a cappuccino with skimmed milk. I was tired because I was on a glucose rollercoaster.

As I've discovered more about glucose, I've learned that there is a wide array of unwelcome short-term symptoms associated with spikes and dips, and they vary from person to person. For some, they're dizziness, nausea, heart palpitations, sweats, food cravings and stress; for others, like me, they're exhaustion and brain fog. And for many Glucose Goddess community members, a glucose spike can also bring on poor mood or anxiety.

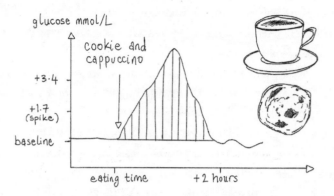

glucose mmol/L

cookie and cappuccino

+3.4

+1.7 (spike)

baseline

eating time        +2 hours

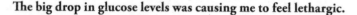

**The big drop in glucose levels was causing me to feel lethargic.**

In the long term, the processes that spikes set into motion – oxidative stress, glycation, inflammation and insulin excess – lead to chronic conditions, from type 2 diabetes to arthritis and depression.

## Short-term effects

### Constant hunger

Are you hungry all the time? You are not alone.

First, many of us feel hungry again shortly after we eat – and here again, it has to do with glucose. If you compare two meals that contain the *same number of calories*, the one that leads to a smaller glucose spike will *keep you feeling full for longer*. Calories aren't everything (more on that in Part III).

Second, constant hunger is a symptom of high insulin levels. When there is a lot of insulin in our body, built up over years of glucose spikes, our hormones get mixed up. *Leptin*, the hormone that tells us we are full and should stop eating, has its signal blocked, while *ghrelin*, the hormone that tells us we are hungry, takes over. Even though we have fat reserves, with lots of energy available, our body tells us we need more – so we eat.

As we eat, we experience more glucose spikes, and insulin rushes in to store excess glucose as fat, which then increases the action of ghrelin. The more weight we put on, the hungrier we get. It's an unfortunate, vicious and unfair cycle.

The answer is not to try to eat less; it is to decrease our insulin levels by flattening our glucose curves – and this often actually means eating more food, as you'll see a bit later. You'll hear the story of Marie, a community member who used to have to eat every 90 minutes and now doesn't even snack any more.

## Cravings

Our understanding of cravings changed thanks to an experiment that took place on the Yale University campus in 2011. Subjects were recruited and placed into an fMRI scanner, which measures brain activity. Then the subjects looked at photos of food on a screen – salad, burger, cookie, broccoli – and rated how much they wanted to eat them on a scale from 1 for 'not at all' to 9 for 'very much'.

On a computer monitor, researchers watched which part of the subjects' brains activated as they looked at the photos. The subjects had also given their consent to be hooked up to a machine that monitored their glucose levels.

What the researchers discovered was fascinating. When the subjects' glucose levels were stable, they didn't rate many of the foods highly. However, *when their glucose levels were decreasing*, two things happened. First, the craving centre of their brain lit up when pictures of high-calorie foods were shown. Second, the participants rated those foods much higher on the 'I want to eat it' scale than when their glucose levels were stable.

The finding? A decrease in glucose levels – even a small decrease of 1.1 mmol/L, which is less than the 1.7 mmol/L dip that occurs after we spike – makes us crave high-calorie foods.

The problem is, our glucose levels decrease all the time – specifically, they drop after every spike. And the higher the spike has been, the more intense the crash will be. That's good, because it means insulin is doing its job, stashing excess glucose in various storage units. But it also means that we're hit by a desire for a cookie or a burger – or both. Flattening our glucose curve leads to fewer cravings.

## Chronic fatigue

Remember your grandfather and his terrible post-retirement gig? When his cabin was stuffed with too much coal, he had to give up shovelling, and the train stopped. The same thing happens to our mitochondria: too much glucose makes them quit, energy production is compromised, and we are tired.

Experiments involving people on stationary bikes show what happens when mitochondria don't work well: those born with mitochondrial defects can typically exercise for only half as long as those without these defects. If you have damaged mitochondria, picking your kid up is more challenging, carrying groceries is exhausting and you won't be able to handle stress (such as a layoff or a breakup) as well as you used to. Mitochondria-generated energy is required to overcome difficult events, whether physical or mental.

When we eat something that tastes sweet, we may think that we are helping our body get energised, but it's just an impression caused by the dopamine rush in our brain that makes us feel high. With every spike, we are impairing the long-term ability of our mitochondria. Diets that cause glucose rollercoasters lead to higher fatigue than those that flatten glucose curves.

## Poor sleep

A common symptom of dysregulated glucose is waking up suddenly in the middle of the night with a pounding heart. Often, it's the result of a glucose crash while we're asleep. Going to bed with high glucose levels or right after a big glucose spike is also associated with insomnia in postmenopausal women and sleep apnoea in a segment of the male population. If you want a good night's sleep, flatten your curves.

## Colds and coronavirus complications

After a glucose spike, your immune system is temporarily faulty. If your glucose levels are chronically elevated, you can say goodbye to five-star immune responses against invaders – you will be more susceptible to infection, and this is particularly so, it turns out, in the case of the coronavirus. Good metabolic health (another way to describe how well your mitochondria are functioning) is one of the main factors that predict whether you survive a coronavirus infection; people with elevated glucose levels have been shown to be more easily infected, to more easily suffer complications, and to be more than twice as likely to die from the virus as people with normal glucose levels (41 per cent versus 16 per cent).

## Gestational diabetes is harder to manage

In every woman, insulin levels increase during pregnancy.

That's because insulin is responsible for encouraging growth – growth of the baby and growth of the mother's breast tissue so she can prepare to breastfeed.

Unfortunately, sometimes this extra insulin can lead to insulin resistance, meaning our body no longer responds to insulin as well as it once did. Our insulin levels rise, but that doesn't help stash excess glucose in the three 'storage lockers' any better, and our glucose levels rise, too. This is what we call gestational diabetes. It's a scary experience for mothers, even more so because it gets worse as the baby's due date approaches.

But by flattening their glucose curves, mothers can reduce their likelihood of needing medication, reduce the birth weight of their baby (which is good because it makes birth easier and is healthier for the baby), and reduce the likelihood of a C-section, as well as limit their own weight gain during pregnancy. This is exactly what Amanda, whom you'll meet in Part III, was able to do.

## Hot flushes and night sweats

As hormone levels drop dramatically in menopause, the changes can feel like an earthquake – everything is thrown off balance, and women experience symptoms from reduced libido and mood swings, to night sweats, insomnia, hot flushes and more.

Research shows that the symptoms of menopause are more pronounced in women who have high glucose and high insulin levels. But there's hope: a 2020 study from Columbia University found that flattening glucose curves is associated with fewer menopause symptoms, such as insomnia.

## Migraine

Migraine is a debilitating condition, which comes in many forms. It's a young field of study, but data proves that women with insulin resistance are twice as likely to have regular migraine headaches as women who don't. When sufferers' insulin levels are lowered, things seem to get better: when treated with a medication that reduced the amount of insulin in the body, over half of a group of 32 people experienced a significant reduction in migraine frequency.

## Memory and cognitive function issues

If you're about to take a test, balance a chequebook, or start an argument that you want to win, beware of what you eat just beforehand. It's easy to reach for something sweet when you want an energy boost, but this choice can affect your brain power. It turns out that big glucose spikes can impair memory and cognitive function.

This effect is worst first thing in the morning, after fasting throughout the night. I wish I had known this growing up, back when I ate a Nutella crepe for breakfast every day. If you have a 9 a.m. meeting in which you want to impress, eat a breakfast that will keep your glucose curve flat. See Hack 4, 'Flatten your breakfast curve', in Part III.

## Type 1 diabetes is harder to manage

Type 1 diabetes is an autoimmune condition in which people lose the ability to make insulin – the cells in their pancreas that control its production don't work.

Every time people with type 1 diabetes experience a glucose spike, their body cannot stash excess glucose in those three storage units because there's no insulin to help. As a result, they need to inject themselves with insulin many times a day to compensate. But large spikes and dips are a daily, and stressful, challenge. By flattening their glucose curves, people with type 1 diabetes can lessen that challenge. Many things can get easier: they can exercise without fear of hypoglycaemia (a state caused by low glucose levels), go to the bathroom less frequently (a side effect of glucose spikes) and even improve their mood.

All the hacks in Part III apply to people with type 1 diabetes as well (and in Hack 10 you'll read a story about Lucy, who has type 1 diabetes and successfully flattened her curves with the hacks). If you have type 1 diabetes, it's important to speak to your doctor before you embark on any dietary changes. Make sure your insulin dosage is adapted if needed.

## Long-term effects

### Acne and other skin conditions

Raise your hand if you wish you had known this in high school: starchy and sugary foods can set off a chain reaction that can show up as acne on your face and body and can even make your skin look visibly redder. This is because many skin conditions (including eczema and psoriasis) are driven by inflammation, which, as you learned, is a consequence of glucose spikes.

When we eat in a way that flattens our glucose curves, acne clears up, pimples get smaller and inflammation is tamed. In a study in males aged 15 to 25, the diet that resulted in the flattest glucose curves led to a significant reduction in acne compared to a diet that caused glucose spikes. (Interestingly, they saw improvements even without reducing other foods known to contribute to acne, such as dairy products.)

## Ageing and arthritis

Depending on your diet, you may have spiked your glucose (and fructose) tens of thousands more times than your neighbour has by the time you reach 60. This will influence not just how old you *look externally* but how old you *are internally*. The more often we spike, the faster we age.

Glycation, free radicals and the subsequent inflammation are responsible for the slow degradation of our cells – what we call *ageing*. Free radicals also damage collagen, the protein found in many of our tissues, which causes sagging skin and wrinkles and can lead to inflammation in joints, rheumatoid arthritis, degradation of cartilage and osteoarthritis: our bones get brittle, our joints are in pain and we definitely can't go for a run in the park.

If there are too many free radicals and too much damage inside a cell, that cell can decide to undergo cell death to prevent further issues. But this isn't without consequences. When cells die, parts of us disappear: our bones waste away, our immune system weakens, our heart pumps less well and neurodegenerative diseases such as Alzheimer's and Parkinson's disease can develop.

Flattening our glucose curves, along with exercising and reducing stress, is a potent way to slow ageing.

## Alzheimer's and dementia

Of all organs, the brain uses the most energy. It's home to a *lot* of mitochondria. That means that when there is excess glucose in our body, our brain is vulnerable to the consequences. The neurons in our brain feel oxidative stress as any other cells do: repeated glucose spikes, because they increase oxidative stress, lead to neuroinflammation and eventually cognitive dysfunction. On top of that, chronic inflammation is a key factor in almost all chronic degenerative diseases, including Alzheimer's.

Indeed, Alzheimer's and glucose levels are so closely connected that Alzheimer's is sometimes called 'type 3 diabetes' or 'diabetes of the brain'. For instance, people with type 2 diabetes are four times as likely to develop Alzheimer's as people without diabetes. The signs are visible early, too: poorly controlled glucose in people with type 2 diabetes is associated with deficits in memory and learning.

Like the other symptoms mentioned here, it's possible that even cognitive decline is reversible: a growing number of studies show short-term and long-term improvements in memory and cognition when patients are put on a glucose-steadying diet. A therapeutic programme out of UCLA found that after just three months of flattening their curves, people who had had to leave their jobs because of cognitive impairment were able to return to work and even perform better than before.

Cancer risk

Children born today have a one in two chance of developing cancer in their lifetime. And poor diet, together with smoking, is the main driver in 50 per cent of cancers.

For starters, research documents that cancer may begin with DNA mutations produced by free radicals. Second, inflammation promotes cancer's proliferation. Finally, when there is more insulin present, cancer spreads even faster. Glucose is the key to many of these processes, and it shows in the data – people with fasting levels higher than 5.5 mmol/L, what is classified as prediabetes, have over double the likelihood of dying of cancer. Flattening glucose and insulin curves is thus an important step to helping prevent the development of cancer.

## Depressive episodes

Your brain doesn't have sensory nerves, so when something is wrong, it can't alert you with pain as other organs do. Instead, you feel mental disturbances – such as poor mood.

When people eat a diet that leads to erratic glucose levels, they report more depressive symptoms and more mood disturbances compared to those on a diet of similar composition but with steadier glucose levels. And the symptoms get worse as the spikes get more extreme, so any effort to flatten the curve, even moderately, could help you feel better.

## Gut issues

It's in our gut that our food is processed, broken down into molecules absorbed into our blood or sent out to the garbage disposal. So it's no surprise that bowel distress – such as leaky gut, irritable bowel syndrome, and slowed intestinal transit – is linked to diet. The jury is still out on the link between glucose spikes and specific digestive issues, but it seems that high glucose levels could increase leaky gut syndrome. Indeed, inflammation – one of the processes set off by glucose spikes – can cause holes in the gut lining, so that toxins that aren't supposed to get through do get through (this is what leads to leaky gut). This in turn leads to food allergies and other autoimmune diseases, such as Crohn's disease and rheumatoid arthritis.

On another note, people who adopt a glucose-flattening diet can get rid of their heartburn or acid reflux very quickly – sometimes within one day.

What's more, we're discovering that gut health is linked to mental health – unhealthy microbiomes can contribute to mood disorders. The gut and the brain are connected by 500 million neurons (that's a lot, but the brain contains a whopping 100 billion). Information is sent back and forth between them all the time, which could be why what we eat, and whether or not we have glucose spikes, affects how we feel.

## Heart disease

When we talk about heart disease, cholesterol is often the main topic of conversation. But that conversation is shifting; we've discovered that it isn't just a matter of 'too much cholesterol'. In fact, half of the people who have a heart attack have *normal* levels of cholesterol. We now know that it's a specific type of cholesterol (LDL pattern B) as well as inflammation that drive heart disease. Scientists have found out why this is the case. And it's linked to glucose, fructose and insulin.

First, glucose and fructose: the lining of our blood vessels is made of cells. These cells are particularly vulnerable to mitochondrial stress – and glucose and fructose spikes lead to oxidative stress. As a result, the cells suffer and lose their smooth shape. The lining of the vessels becomes bumpy, and fat particles get stuck more easily along the uneven surface.

Second, insulin: when our levels of insulin are too high, our liver starts producing LDL pattern B. This is a small, dense kind of cholesterol that creeps along the edges of the vessels, where it's likely to get caught. (LDL pattern A is large, buoyant and harmless – we get it from eating dietary fat.)

Finally, if and when that cholesterol is oxidised – which happens the more glucose, fructose and insulin are present – it lodges under the lining of our blood vessels and sticks there. Plaque builds up and obstructs the flow, and this is how heart disease starts.

Spikes drive these three processes. That's why science is finding that even if our fasting glucose is normal, each additional glucose spike increases our risk of dying of a heart attack. To help our heart, we should flatten our glucose, fructose and insulin curves.

Nine out of ten doctors still measure *total* LDL cholesterol to diagnose heart disease and prescribe statins if it's too high. But what's important is LDL pattern B and inflammation. To add to the problem, statins lower LDL pattern A, but they don't lower the problematic pattern B. This is why statins don't decrease the risk of a first heart attack.

Here again, glucose and fructose and the inflammation that high levels of these molecules cause in our bodies, are the key to understanding this disease. Doctors can better measure heart disease risk by looking at what's called the triglycerides-to-HDL ratio (which tells us about the presence of the small, dense LDL pattern B), and C-reactive protein (which tells us about inflammation levels). Triglycerides become LDL pattern B in our bodies. So by measuring triglycerides, we can gauge the amount of the problematic LDL pattern B in our system. If you divide the level of triglycerides (in mg/dL) by HDL level (in mg/dL), you'll get a ratio that is surprisingly accurate in predicting LDL size. If the result is smaller than 2, that's ideal. If the result is above 2, it can be problematic. Then, because inflammation is a key driver of heart disease, measuring C-reactive protein, which increases as inflammation does, is better at predicting it than cholesterol levels.

### Infertility and polycystic ovary syndrome (PCOS)

Scientists have recently discovered a remarkable connection between insulin and reproductive health. It turns out that insulin levels are an important piece of information used by the brain and your gonads, or sex organs, to decide whether your body is a safe environment in which to conceive. If your insulin is out of whack, your body isn't too keen on reproducing, because it suggests that you aren't healthy. Both women and men with high insulin levels are more likely to be infertile. The more glucose spikes in our diet, the higher our insulin levels and the higher our incidence of infertility.

When it comes to female infertility, PCOS is often to blame. One in eight women experience it, and when they do, their ovaries become burdened with cysts and no longer ovulate.

PCOS is a disease caused by too much insulin. The more insulin that is present, the more PCOS symptoms. Why? Because insulin tells the ovaries to produce more testosterone (the male sex hormone). On top of that, with too much insulin, the natural conversion from male to female hormones that usually takes place is hampered – which leads to even more testosterone in the body. Because of the excess testosterone, women suffering from PCOS display masculine traits: hair in places where they don't want hair (such as the chin), baldness, irregular or missed periods, or acne. Ovaries can also retain and accumulate eggs, stopping ovulation.

Many women with PCOS also have a hard time losing weight – because where there is too much insulin, there is an inability to burn fat.

Some women are more susceptible to PCOS than others (not every woman with high insulin levels has PCOS), but in all cases, keeping glucose levels under control can reduce and even completely alleviate all symptoms. In Part III, you'll meet Ghadeer, who rid herself of PCOS symptoms, reversed her insulin resistance and lost over 20 pounds using the hacks in this book. In a study performed at Duke University, women who went on a glucose curve-flattening diet for six months cut their insulin levels by half and, as a consequence, their testosterone levels by 25 per cent. Their body weight dropped and their body hair diminished as their hormones came into balance, and two of the 12 participants became pregnant in the course of the study.

For men, dysregulated glucose is also linked to infertility: elevated glucose levels are associated with reduced quality of semen (fewer viable candidates) and erectile dysfunction – so much so that recent studies suggest that erectile dysfunction in men under 40 years old could be due to an unknown metabolic and glucose dysregulation issue. If you are trying to have a baby, flattening your glucose curves is a big help.

### Insulin resistance and type 2 diabetes

Type 2 diabetes is a global epidemic, with half a billion people in the world suffering from the disease and the number rising every year. It's also the best-known health condition associated with elevated glucose levels. To better demonstrate how spikes lead to type 2 diabetes and how to reverse the condition, let me tell you a story about my espresso habit.

When I was a student in London, I steadily increased my daily dose of coffee. I started with one espresso in the morning and after a few years somehow ended up at five a day just to stay awake. I had to keep upping my dose of caffeine to feel the same effect as before. In other words, I gradually became *resistant* to caffeine.

It's the same with insulin. When insulin levels have been high for a long time, our cells start becoming resistant to it. Insulin resistance is the root cause of type 2 diabetes: liver, muscle and fat cells need larger and larger quantities of insulin to take up the same amount of glucose. Eventually, the system doesn't work any more. Glucose is no longer stored away as glycogen, even though our pancreas produces growing quantities of insulin. The result is that the glucose levels in our body are increased for good. As our insulin resistance gets worse, we go from prediabetes (fasting glucose levels above 5.5 mmol/L) to type 2 diabetes (above 7.0 mmol/L). Slowly but surely, over many years, every glucose spike you experience will contribute to worsening your insulin resistance and raising the overall baseline glucose level in your body.

The common (but misguided) method of treating type 2 diabetes is to give the patient more insulin. This brings the glucose levels down temporarily by forcing the fat cells – that large storage unit – to open (and make them put on weight).

A vicious cycle is created, where higher and higher doses of insulin are administered and the patient's weight goes up and up, but the root problem of high insulin levels isn't addressed. Adding extra insulin helps people with type 2 diabetes in the short term, bringing their levels down after eating, but in the long term worsens the condition.

What's more, we now know that type 2 diabetes is an inflammatory disease – more inflammation, a process set off by glucose spikes, makes it worse.

It makes sense, therefore, that a diet that reduces our intake of glucose and consequently our production of insulin, would help reverse type 2 diabetes. A 2021 review of 23 clinical trials made it clear that the most effective way to reverse type 2 diabetes is to flatten our glucose curves. This is more effective than low-calorie or low-fat diets, for example (even though they can also work). In one study, people with type 2 diabetes who changed their diet and reduced their glucose spikes cut their insulin injections by *half* within one day. (If you are on medication, speak to your doctor before you try the hacks in this book – as you see, changes can be very rapid.)

In 2019, the American Diabetes Association (the ADA) started endorsing glucose-flattening diets in the light of mounting compelling evidence that following these improves type 2 diabetes outcomes. In Part III, you'll learn how to flatten your glucose curves while still eating what you love.

## Non-alcoholic fatty liver disease

Liver disease used to be a problem only for those who drank a lot of alcohol.

But in the 21st century, that changed. Robert Lustig, an endocrinologist, was confronted with a startling fact in his practice in San Francisco in the late 2000s: some of his patients were showing signs of liver disease, but they weren't heavy drinkers. In fact, many of them were under the age of ten.

He went on to discover that excess fructose could cause liver disease, just as alcohol does. To protect us from fructose, just as it does with alcohol, the liver turns it into fat, thereby removing it from the bloodstream. But when we repeatedly eat things high in fructose, our liver itself becomes *fatty* – which happens with alcohol, too.

The medical community named this new condition non-alcoholic fatty liver disease (NAFLD) or non-alcoholic steatohepatitis (NASH). It's extremely common: around the world, one out of every four adults has NAFLD. In people who are overweight, it's even more common: over 70 per cent of them have it. Unfortunately, this condition can get worse over time, leading to liver failure or even cancer.

To reverse the condition, the liver needs a break so that it can deplete its excess fat reserves. The way to do this is by lowering our fructose levels and preventing further fructose spikes – which happens naturally when we flatten our glucose curves (because fructose and glucose go hand in hand in food).

## Wrinkles and cataracts

Do you know why some 60-year-olds look as though they are 70, while others look as though they are 45? That's because we can influence the speed at which we age – and one of the ways to do it is by flattening our glucose curves.

Glucose spikes, as I explained in the previous chapter, result in glycation – and glycation makes us age faster and look older.

For example, when glycation transforms a molecule of collagen, it makes it less flexible. Collagen is needed to repair wounds, as well as make healthy skin, nails and hair. Broken collagen leads to sagging and wrinkling. The more glycation, the more sagging skin and wrinkles. Crazy but true.

Glycation happens everywhere in our body; when it occurs in our eyes, the damaged molecules start to clump together and, over time, form cataracts, blocking out the light.

Science, including the research I've shared here, helps you decode the messages from your body. Take a second and check in. How are you feeling? What parts hurt? Which systems feel sluggish? If you could wake up every day and feel amazing, wouldn't you want to?

Odds are, you are among the 88 per cent of adults who have dysregulated glucose levels and experience, without knowing it, the many consequences of spikes I've just described – from short-term side effects to long-term illnesses. From wrinkles and acne to cravings, hunger, migraine and depression, to poor sleep, infertility and type 2 diabetes, these symptoms are messages from your body. And although these issues are very common, recent discoveries show us that they are also very much reversible.

In Part III, I show you how to start that process. You're

about to discover food hacks that will help you flatten your glucose curves, reconnect with your body and reverse your symptoms – while still eating what you love. I hope that soon you'll wake up one morning and feel amazing. Because that's exactly what happened to Bernadette, whom you're about to meet.

# PART III

# HOW CAN I FLATTEN MY

# GLUCOSE CURVES?

# HACK 1:
## EAT FOODS IN THE RIGHT ORDER

'I lost five pounds in nine days,' Bernadette told me on a sunny Tuesday morning, 'and all I did was change the order in which I eat my food.'

So often, we focus on *what* and *what not* to eat. But what about *how* to eat? It turns out that *how* we eat our food has a powerful effect on our glucose curves.

Two meals consisting of the same foods (and therefore the same nutrients and the same calories) can have vastly different impacts on our body depending on *how* their components are eaten. I was taken aback when I read the scientific papers that proved this, notably a seminal one out of Cornell University in 2015: if you eat the items of a meal containing starch, fibre, sugar, protein and fat in a specific order, you reduce your overall glucose spike by *73 per cent*, as well as your insulin spike by *48 per cent*. This is true for anyone, with or without diabetes.

What is the right order? It's fibre first, protein and fat second, starches and sugars last. According to the researchers, the effect of this sequencing is *comparable to the effects of diabetes medications* that are prescribed to lower glucose spikes. A startling study from 2016 proved the finding even more definitively: two groups with type 2 diabetes were given a standardised diet for eight weeks and asked to either eat their food in the right order or eat it however they pleased. The group who ate their food in the right order saw a significant reduction in their HbA1c level, which means they started reversing their type 2 diabetes. The other group, eating the exact same food and number of calories but in no particular order, didn't see an improvement in their condition.

Talk about a groundbreaking discovery.

The explanation for this surprising effect has to do with how our digestive system works. In order to visualise it, think of your stomach as a sink and your small intestine as the pipe below it.

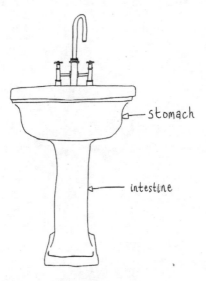

**Picture your stomach as a sink and your intestine as the pipe below it.**

Anything you eat lands in your sink, then flows through to your pipe, where it is broken down and absorbed into your bloodstream. Every minute, on average, about three calories' worth of food trickle through from sink to pipe. (This process is called *gastric emptying*.)

If starches or sugars are the first thing to hit your stomach, they get to your small intestine very quickly. There, they are broken down into glucose molecules, which then make it through to the bloodstream very quickly. That creates a glucose spike. The more carbs you eat and the quicker you eat them, the more forcefully the load of glucose appears – the bigger the glucose spike.

Say you have both pasta and vegetables on your plate (broccoli, anyone? I love broccoli) and you eat the pasta first, then the broccoli. The pasta, which is a starch, turns into glucose as it is quickly digested. The broccoli then 'sits' on top of the pasta and waits its turn to go through the pipe.

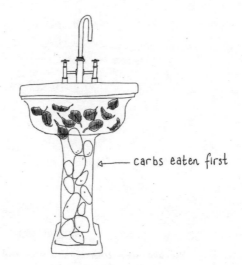

carbs eaten first

**When you eat carbs first, they flow into your intestine uninterrupted.**

However, consuming the *veggies first* and the *carbs second* significantly changes what happens.

Begin by munching on the broccoli. Broccoli is a

vegetable, and vegetables contain plenty of fibre. As we've seen, fibre isn't broken down into glucose by our digestive system. Instead, it goes through from sink to pipe to… sewage, slowly and unchanged. But that's not all.

Fibre has three superpowers: first, it reduces the action of alpha-amylase, the enzyme that breaks starch down into glucose molecules. Second, it slows down gastric emptying: when fibre is present, food trickles from sink to pipe more slowly. Finally, it creates a viscous mesh in the small intestine; this mesh makes it harder for glucose to make it through to the bloodstream. Through these mechanisms, fibre slows down the *breakdown* and *absorption* of any glucose that lands in the sink after it; the result is that fibre flattens our glucose curves.

Any starch or sugar that we eat *after fibre* will have a reduced effect on our body. We'll get the same pleasure from eating it but with fewer consequences.

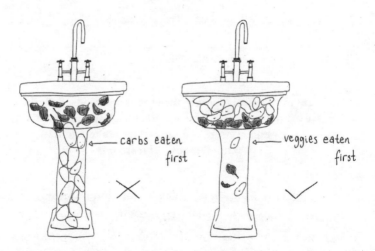

**Eating veggies first and carbs second greatly slows down the speed at which glucose makes it to the bloodstream, thereby flattening the glucose spike associated with that meal.**

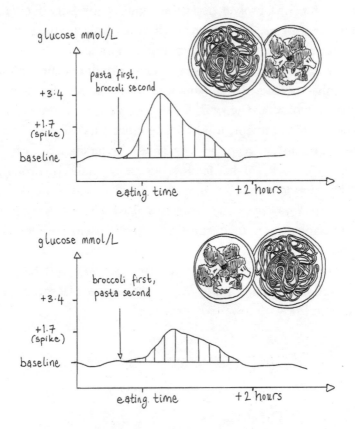

**These two meals contain the exact same foods. But when we eat the vegetables first and the starch second, we flatten our glucose curve and experience fewer, and lesser, side effects of a glucose spike.**

That's carbs and vegetables. Now enter protein and fat. Protein is found in meat, fish, eggs, dairy, nuts, beans and legumes. Foods that contain protein often contain fat, too, and fat is also found on its own in foods such as butter, oils and avocados. (Incidentally, there are good and bad fats, and the bad fat that we should avoid is found in hydrogenated and refined cooking oils made from rapeseed, corn, cotton-seed, soybean, safflower, sunflower and grapeseed.) Foods containing fat also slow down gastric emptying, so eating them *before* rather than *after* carbs also helps flatten our glucose curves. The takeaway? Eating carbs after everything else is the best move.

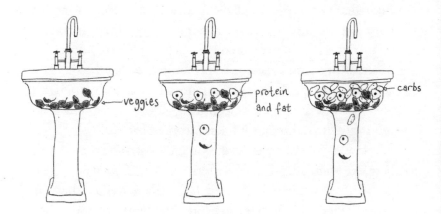

**The right order to eat foods in: veggies first, protein and fat second, starch last.**

To illustrate the effect of food order on glucose spikes, return to the Tetris analogy: blocks coming down slowly are easier to arrange than blocks coming down quickly. When we eat foods in the right order – veggies first, protein and fats second, carbs last – not only do we slow down the *speed* of the blocks, we even cut down on the *quantity* of blocks thanks to the mesh that fibre adds to our intestine. The slower the trickling of glucose into our bloodstream, the flatter our glucose curves and the better we feel. We can eat *exactly the same thing* – but by eating carbs last, we make a big difference in our physical and mental well-being.

What's more, when we eat foods in the right order, our

pancreas produces less insulin. And as I explained in Part II, less insulin helps us return to fat-burning mode more quickly, the results of which include losing weight.

## Meet Bernadette

Bernadette – who doesn't have diabetes – had been using this hack not because she wanted to lose weight (her girlfriends had warned her that postmenopausal pounds were impossible to shed) but simply because she wanted to feel better. Her attempt at weight loss had come to a halt a few years back. She was sick and tired of counting calories. She had tried intermittent fasting but it hadn't worked for her.

Now, at 57 years old, what bothered Bernadette most was her poor energy level. Every afternoon, like clockwork, she'd get so tired while going about her daily activities that she would glance at the floor at work, the bank, or the coffee shop, and think, *If only I could lie down there, I would have a fabulous nap.* To make it through the afternoon, she'd eat chocolate bars. But when it came to going to sleep at night, she suffered from insomnia, waking up around four every morning.

Bernadette first learned about glucose spikes on the Glucose Goddess Instagram account. She didn't know whether she was in fact experiencing spikes, but she decided to try this hack to see if it might help her. When she found herself in her kitchen the next day at lunchtime, with the ingredients for her usual sandwich on her countertop, she recalled the 'veggies first, protein and fat second, carbs last' hack and thought, *Hm. Instead of stacking everything and eating the sandwich as one, I could eat the salad and pickle first, then the tuna, then the toasted bread.* She placed each on her plate and ate her newly baptised 'deconstructed sandwich'.

Bernadette is a creature of habit, and her go-to dinner is steak with veggies and pasta. So that day, she had the veggies and meat first and the pasta last. At no point did she change the quantity of food she was consuming – just the order in which she ate it.

The next day, to her great surprise, she woke up feeling rested for the first time in months. When she reached for her phone to check the time, she saw 7 a.m. – hours later than when she usually opened her eyes. I know this sounds crazy – Bernadette thought it was crazy, too. But she was thrilled. So she kept going, deconstructing her sandwiches and eating her pasta last in the evenings.

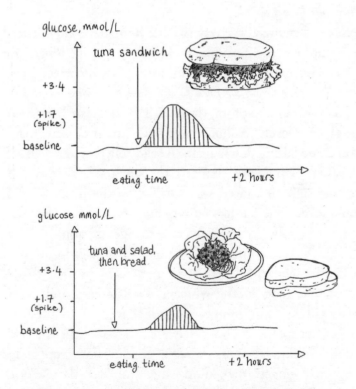

**Deconstruct a sandwich and eat the bread (starch) last to dampen the glucose spike it creates and get rid of the 3p.m. sleepiness that comes when your glucose levels plummet.**

After three days, her longings for a mid-afternoon nap were gone. She was energised. She felt better than she had in years. When she went to the supermarket next, instead of stocking up on chocolate bars as she usually did, she didn't feel the need to buy any. 'It was so freeing,' she said.

> TRY THIS: Next time you sit down for a meal, eat the veggies and proteins first and the carbs last. Note how you feel after eating compared to how you usually feel after a meal.

### What was going on?

Before she changed the way she ate, Bernadette experienced the symptoms of a post-lunch glucose crash. She longed for a nap. Her brain sent a well-meaning, but incorrect, alert: *We're low on energy, we need to eat something.* She looked for a chocolate bar and promptly ate it. The chocolate bar caused her glucose levels to shoot right back up, after which they soon came falling down again. A wild rollercoaster ride.

When Bernadette changed the order of the food she ate, the spike it caused was smaller, so the drop was less pronounced. She felt less hungry and less tired in the afternoons. The rollercoaster ride came to a gentle stop.

There's a scientific explanation for this improvement in her hunger: the Cornell research team showed that if we eat our food in the wrong order (starches and sugars first), ghrelin, our hunger hormone, returns to pre-meal levels after just two hours. If we eat our food in the right order (starches and sugars last), ghrelin stays suppressed for much longer (they didn't measure past three hours, but looking at the trends, I think it fair to say that it stays down for five to six hours).

Research also shows that in postmenopausal women, a diet with fewer glucose spikes is associated with a lower incidence of insomnia. What's more, when we sleep better, we make better choices, and it's easier to find the motivation to do good for ourselves. Bernadette felt that way – she even started going for walks in the afternoon.

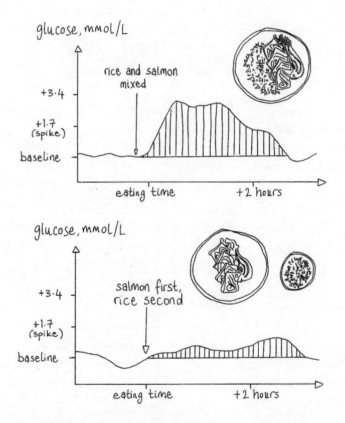

**Even if there are no veggies on our plate, 'deconstructing' our meals and eating carbs last helps our body. We significantly flatten the glucose curve and reduce the likelihood of weight gain, cravings, lethargy and the harmful long-term side effects of elevated glucose levels.**

Nine days into what felt like the easiest lifestyle change she had ever tried, Bernadette's jeans started feeling looser. So she jumped on the scales. To her surprise, she was down 5 pounds. In just over a week, she had lost almost a third of the weight she had put on since menopause without even trying.

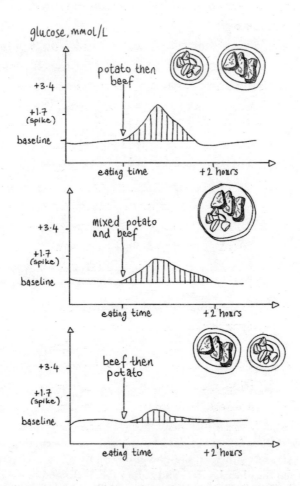

**Eating the potato first led to the biggest spike, mixing it with the meat together was better, but starting with the meat and saving the carbs for last was best for my glucose levels.**

Remember, in the cockpit of our body, getting our glucose lever into the right position is the most powerful thing we can do. The consequences are often surprising, such as unintended weight loss. And as you see, it starts with something as easy as eating in the right order.

### I thought fruit was supposed to be eaten alone, otherwise it rots in our stomach?

A question I am often asked when I talk about this hack involves fruit. I categorise fruits in the 'sugars' category, because although they contain fibre, they are made up mostly of glucose, fructose and sucrose – aka sugars. Therefore they should be eaten last. But people ask, 'Doesn't eating fruit last cause it to rot in the stomach?' The short answer is no.

This false belief seems to date back to the Renaissance, around the time the printing press was invented. Some doctors at the time recommended that you should *never* finish a meal with raw fruit, because it would 'float on top of the contents of the stomach and eventually putrefy, sending noxious vapours into the brain and disrupting the entire bodily system'.

As it turns out, there is no evidence supporting this.

Rotting happens when bacteria lodge on food and start digesting that food to fuel their own growth. The white and green specks you see on a strawberry you've left too long in the fridge are bacteria growing. First off, rotting takes days or weeks to happen. It can't happen in a few hours, which is about how long it takes for fruit to be digested.

Second, our stomach is an acidic environment (pH 1–2), and any environment with a pH below 4 prevents bacterial overgrowth (and therefore rotting). Nothing can rot in the stomach, and in fact, the stomach, together with the oesophagus, is the place where there are the *fewest* bacteria in our entire digestive tract.

Those Renaissance physicians didn't have it right. But there are many instances of cultures throughout history that have embraced the 'right food order': in Roman times, a meal generally started with eggs and ended with fruit. During the Middle Ages in Europe, banquets usually ended with fruit to 'close up digestion'. Today, people in most countries end meals on a sweet note: dessert.

To be fair, maybe the doctors of the 1300s were not completely off the wall when they recommended that you eat fruit alone. A handful of people have shared with me that they have to eat fruit on its own; otherwise they experience discomfort, such as bloating or gas. It all comes back to listening to our body. Starches and sugars last are the right way to go, unless we personally feel that it doesn't agree with us.

### How quickly can I eat the foods one after the other?

Many different timings were studied in clinical settings – 0 minutes, 10 minutes, 20 minutes; they all seem to work. As long as you eat starch and sugars last, even if it's without stopping, you will flatten your glucose curve. During my meals, I just eat one food group right after the other (and so does Bernadette).

### What if there aren't any starches or sugars in the meal?

Naturally, a meal with no starches or sugars will lead to only a very moderate glucose spike (some protein turns to glucose as well, but at a much lower rate than carbohydrates). However, it is still beneficial to start with veggies and have protein and fats second.

### Do I have to do this all the time?

It's up to you to use the hacks in this book in a way that

makes sense for you. In my own life, I eat foods in the right order when it's easy. If I'm eating a dish such as curry or paella, where veggies, proteins, fats and carbs are mixed together and it's difficult to separate the ingredients, I don't stress about it. I sometimes have a few bites of veggies first, then eat the rest of the dish mixed.

The most important thing to remember is that it's best to eat starches and sugars *as late in the meal as possible*. And remember to celebrate the small changes: if you eat your veggies first, then mix starches with your protein and fat, that's still an improvement and is better than eating the veggies last.

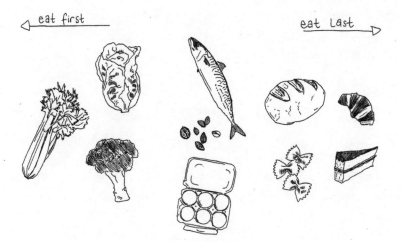

## Let's recap

Whenever it's feasible and doesn't turn your meal into a complicated ordeal involving painstakingly separating the elements of the chef's special, it's best to eat anything that turns into glucose last. Start with the veggies and greens on your plate, then fat and protein, then starches and sugars. It's tempting to go straight for the carbs when you are hungry, but if you use this hack, your cravings will be curbed later on.

Based on the science, I love any meal that starts with a salad. Unfortunately, many dining experiences don't set us up for success: restaurants serve bread while you're waiting for food. Starting with starch is the absolute opposite of what you ought to do. It will lead to a glucose spike that you won't be able to tame, then a crash later on – which will intensify your cravings.

Now that I think of it, if I had to devise a way to get people to eat more at my restaurant, giving them the bread first is exactly what I would do.

# HACK 2: ADD A GREEN STARTER TO ALL YOUR MEALS

I know what you're probably thinking as you read the title above: *This is the same as the previous hack, eat your veggies first.* No! This hack is on another level. I'm talking about *adding* a dish at the beginning of your meals. You will eat more than you usually do and flatten your glucose curves in the process (and in the next hack, we'll go over why adding these calories is good). The goal here is to return to how food used to be, before it was processed: wherever there were starches and sugars, there was also fibre. By adding a delicious green starter, we're bringing fibre back.

## Meet Jass

A few years ago, I finally got my mother the present she had always wanted: a card that read 'OMG – My mom was right about everything!'

To be fair, she wasn't right about starting the day with Special K and orange juice. But she *was* right about some other things, such as the importance of organising my mail, not buying clothes that needed to be dry cleaned because I'd never get around to dropping them off, and scrubbing the inside of the fridge once a month. But when I first left home and went to college, I didn't follow any of that advice. I sure as heck didn't clean the inside of any kitchen appliances.

As we age, we often realise the wisdom of our parents' advice. As I've studied the science behind glucose spikes, I've seen multiple studies showing that some recommendations for flattening our glucose curves are ones that were encouraged by the previous generation.

That's what Jass discovered, too.

Jass (short for Jassmin) grew up in the countryside of Sweden with a Lebanese mother and a Swedish father. Her parents were busy: they had full-time jobs and five children. But no matter how busy they were, the family sat down to eat together every night. The first course of each dinner was always a large salad.

When Jass moved out and got her first job as a teacher in Gothenburg, just like me, she didn't think to follow her family's example. Her days were paced by the back-and-forths between her apartment and the middle school. She was swamped by coursework and, in between deadlines, tried to maintain her social life – in short, she didn't have time to think about food. Her usual tactic was to swing by the grocery store on her way home from work, pick up a box of pasta and eat that for dinner. She'd pack the leftovers for lunch the next day.

Before she knew it, her eating habits had completely changed. Once someone who just enjoyed chocolate for dessert, she had now developed a sweet tooth. She counted the seconds until break when she could go to the coffee shop and buy a slice of cake. She needed a regular supply of treats to get through the day. The new job was taxing, she worked a lot and was quite tired, and eating something sweet every few hours kept her motivated.

As the months went on, her sweet tooth grew even more pronounced. She was either eating sugar or thinking about eating sugar. Her cravings were out of control. And in fact, her cravings were controlling her. Her willpower was nowhere to be found. She started to gain weight. Acne appeared on her forehead. Her periods became irregular. She felt bad – about the cravings and all the changes taking place in her brain and her body.

One afternoon, before Jass's usual snack time, she asked her class to open their biology books to Chapter 10, 'Metabolism'. She spoke about how our body gets energy from food and in particular what happens when we eat carbohydrates. Jass was teaching a lesson on glucose.

As she went through the material, she couldn't help but think that maybe there was something that could help her here. In the same week, serendipitously, a colleague showed her the Glucose Goddess Instagram account. Things started clicking. She wondered, *Is glucose the problem? Am I experiencing glucose spikes without knowing it? Is this why I can't stop eating chocolate and why I feel so tired all the time?*

She soon noticed two things: (1) when she was hungry, she always went for carbs first and (2) her meals weren't balanced: lunch and dinner were mostly starches. She realised she was getting messages from her body: something was off. Yes, she was definitely on a glucose rollercoaster.

In order to flatten our glucose curves, eating fibre, protein and fat before starchy foods is key, as we have seen. With this empowering realisation, Jass decided to reinstate a tradition from back home: a first course of a large salad every night. She had grown up eating fattoush, a traditional Lebanese salad. So off she went to make it herself: she combined chopped bell peppers, cucumbers, tomatoes and radishes with lettuce, a handful of parsley, spring onions, and seasoned it with olive oil, salt and a lot of lemon juice.

# The more fibre, the better

On average, people in the UK don't get nearly enough fibre: the daily consumption is on average *half* of what it should be. In the US, only *5 per cent* of people meet the recommended daily amount: 25 grams per day. The US government calls it a 'nutrient of public health concern'. This disappearance is due mostly to food processing, as I explained in Part I.

Fibre is in the structural lining of plants – it's abundant in their leaves and bark. So unless you're a wood-eating termite (in which case, I'm impressed that you can read!), you get most of your fibre from beans, vegetables and fruit.

This plant-made substance is incredibly important to us: it fuels the good bacteria in our gut, strengthens our microbiome, lowers our cholesterol levels and makes sure everything runs smoothly. One of the reasons a diet high in fruit and vegetables is healthy is because of the fibre it provides.

As mentioned in the last chapter, fibre is also good for our glucose levels for several reasons, notably because it creates a viscous mesh in our intestine. The mesh slows down and reduces the absorption of molecules from food across the intestinal lining. What does this mean for our glucose curves? First, we absorb fewer calories (we'll talk about calories in the next hack). Second, with fibre in our system, any absorption of glucose or fructose molecules is lessened.

This has been shown many times in scientific settings. For instance, in a 2015 study, scientists in New Zealand fed participants two types of bread: regular bread and bread enriched with 10 grams of fibre per serving. They found that the additional fibre reduced the glucose spike of the bread by over 35 per cent. Talking about bread, here's what you're looking for if you want to enjoy some while flattening your curves: skip the loaves that claim to contain 'whole grain', which often don't have much more fibre than their traditional 'white' counterparts. Buy bread that is dark and dense, made from rye with a sourdough starter. It's traditionally German and usually called *seed bread* or *pumpernickel*. That contains the most fibre.

**Want bread with beneficial fibre in it?**
**Go for the Germanic ones.**

However, even this dark bread is not the best way to add fibre to our diet, as bread contains starch and adds glucose to the meal. You know what's a better way to get fibre? Eating green veggies. These contain mostly fibre and very little starch.

We know that consuming more fibre is beneficial and that eating it *before* all the other foods is even more so (see previous hack). That's why adding a green starter to each of our meals has a powerful effect on our glucose curves.

How big should this green starter be? As big as you like. I've found that the sweet spot is a one-to-one ratio to the carbs you'll eat after. My favourite: two cups of spinach, five jarred artichoke hearts, vinegar and olive oil. My little brother's go-to: one big raw carrot, sliced, with hummus (not technically *green* but still vegetable based, which is what we're looking for). You'll find more ideas later in this chapter.

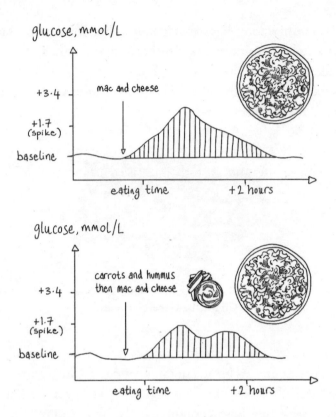

**We can use any types of vegetables as our starter. This includes vegetables that aren't green, such as carrots. You can throw in some pulses, too, such as hummus or lentils, because they are full of fibre as well.**

Around the world, tradition mirrors science: in Iran and the countries of Central Asia, meals start with fresh herbs eaten by the bunch. In the Mediterranean, meals often open with vegetables – marinated aubergines and artichokes for antipasti in Italy, sliced radishes, spring beans and endive as crudités in France, or the combination of finely chopped parsley with ripe tomatoes and cucumbers that makes up tabbouleh from Turkey to Lebanon and Israel. Adding a green starter flattens our glucose curve. With a flatter curve, we stay full for longer and avoid the glucose dip that leads to cravings a few hours later.

So back to Jass.

Jass added fattoush as a starter to her dinner each night. She still ate her usual bowl of pasta after, but now something different was happening in her body: she went from a forceful delivery of glucose to a gentle one. The spike was less pronounced, and the crash that followed was smaller.

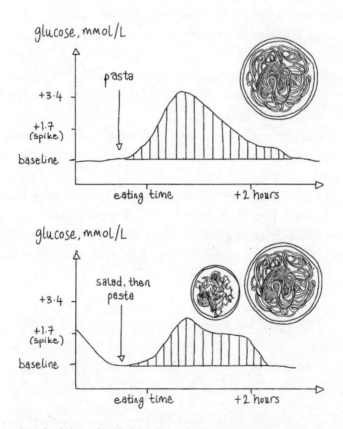

**Jass hadn't realised that when she ate pasta on its own, she got onto a glucose rollercoaster. By adding a salad at the beginning of every meal, she flattened her glucose curve. Her uncontrollable cravings lessened, and her willpower returned.**

Pretty quickly, Jass began to feel better. First, and most noticeable to her, she was able to go longer without eating. After lunch, she felt full until 5p.m., instead of feeling starving at 3p.m. She was more alert. She had more patience with her students. She found herself skipping through the hallways and smiling at her colleagues. The flatter curve had steadied both her hunger and her mood.

About 10 days in, Jass lost her taste for treats. To her great surprise, on her coffee break, she could walk by the local bakery, and think, *Oh, a tasty-looking cake*, without feeling the urge to eat it. There was the *habit* of eating sweets but no torturous impulse to act on it. She didn't need to spend energy trying to suppress her cravings any more – because her cravings were gone. She regained her willpower – in fact, it now felt like a superpower.

When we flatten our glucose curves, the side effects are usually pleasant and unexpected. As in the case of Bernadette, Jass lost weight without trying. So far she's lost 20 pounds, going from 183 pounds to 163 pounds. 'I just cared about keeping my body in a happy, steady glucose zone. Everything else fell into place.' She told me that her period is back to normal, her acne has cleared, her sleep has improved, and she feels better, too.

---

TRY THIS: Think of your favourite veggie or salad. Prepare it with care, and eat it before lunch and dinner for a week. Notice your cravings and whether they change.

---

### How much time do I need to wait between the starter and the main dish?

You don't need to wait at all; you can eat them in sequence. If you do wait, try not to leave more than a couple of hours between your green starter and the rest of your meal. That's because two hours is around the time it takes for fibre to go through the stomach and the top part of the small intestine. For example, if you have a salad at noon and rice at 1p.m., the fibre from the salad will still help flatten the rice's spike. But if you have a salad at noon and rice at 3p.m., it won't.

### How many veggies do I need to eat?
First off, any amount is better than none, and the more, the better. Studies haven't been done on the ideal ratio. But I try to eat the same quantity of veggies as the starches that follow.

If I don't have time to make a salad, I grab two jarred hearts of palm or a couple of pieces of roasted cauliflower that I keep in the fridge. Though that's not a one-to-one ratio, it's enough to start seeing a small benefit, and it's better than having no veggies before a meal at all.

### What qualifies as a green starter?
Any vegetable qualifies, from roasted asparagus to coleslaw, grilled courgettes and grated carrots. We're talking artichokes, aubergines, broccoli, Brussels sprouts, lettuce, pea shoots, rocket and tomatoes, and also pulses, beans and viscous foods such as natto (a Japanese food made from soybeans) – the more, the better.

Incidentally, you can eat them either raw or cooked. But skip juiced or mashed preparations, because the fibre in them is either missing (in the case of a juice) or blitzed to oblivion (in the case of a mash). Soup is another story. Do you remember my answer to my mother when she calls me from the grocery store to ask if a food is 'good' or 'bad'? The answer is relative – and soup is a perfect example of this. Soup is a great dish – it contains plenty of nutrients and vitamins, it's filling and it's one of the healthiest starters you can order at a restaurant. But it's not healthier than eating a whole vegetable. Beware of store-bought soups, too: they are often mostly potatoes, which break down into starch.

### What's the easiest thing to start with?

Buy a bag of spinach at the supermarket, toss 3 cups of it in a bowl with 2 tablespoons of olive oil, 1 tablespoon of vinegar (any kind you like), and salt and pepper, and top with a handful of crumbled feta cheese and toasted nuts. (It's okay, in fact good, to mix some protein and fats into your green starter.) You can also add pesto, grated Parmesan cheese, and some toasted seeds, as you prefer. It should be something quick that you find tasty. This isn't cooking; it's assembling.

Beware of ready-made dressings, as they're often packed with sugar and loaded with vegetable oil – it's better to make a simple one from scratch with the ratio of oil and vinegar I describe above. I make a batch of dressing every Sunday and keep it in the fridge to use all week.

Here are some even quicker things to eat:

- A couple of pieces of leftover roasted veggies (top tip: I often roast a batch of broccoli or cauliflower and keep it in the fridge)
- A few mouthfuls of pickled vegetables
- A sliced cucumber with guacamole
- A sliced tomato with one or two slices of mozzarella cheese
- Baby carrots with hummus
- Four marinated artichokes from a jar
- Two canned hearts of palm
- Two spears of jarred white asparagus

### What do I do at a restaurant?

If my party is ordering starters, I order a salad. If we aren't ordering starters, I ask for a vegetable-based side with my main (such as a simple green salad with olive oil and vinegar, steamed green beans or sautéed spinach), and I eat it before the rest of my dish. I wait until after eating my veggies to eat my main or touch the bread.

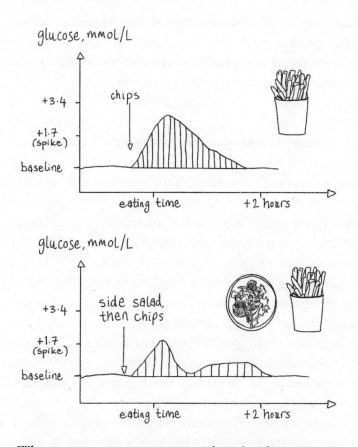

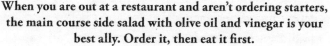

**When you are out at a restaurant and aren't ordering starters, the main course side salad with olive oil and vinegar is your best ally. Order it, then eat it first.**

**What about calories?**
Great question. More on that in the next hack. Stay tuned.

**How about supplements?**
It is always best to eat whole food rather than supplements, but if it's easier on some occasions, a fibre supplement at the beginning of a meal can help.

**Doesn't adding fat (from the salad dressing) to carbs lead to weight gain?**
It doesn't – that's a myth that's been debunked. More on that in Hack 10, 'Put some clothes on your carbs'.

## Meet Gustavo and his wingman, broccoli

People around the world get creative when they use these hacks in their daily lives. Depending on the country and what is available, their interpretations always impress me. I'd like to mention one example of how this hack helped Gustavo, because I find it particularly useful.

Gustavo is a salesman in Mexico. At 50, he has already lost two people close to him to the same disease: his father passed away from type 2 diabetes; then his colleague – years younger – also died of diabetes complications. That was a wake-up call. Gustavo didn't want his life to end because of poor health; he wanted to be an active member of his community for many years to come.

Gustavo hadn't (yet) been diagnosed with diabetes, but he knew he was seriously overweight, and, when he learned that people can experience spikes for years before developing the condition, he was pretty sure he was on his way there, just like his dad.

That said, he also learned that diabetes isn't just about

genes: even if our parents get diabetes, this does not mean that we automatically get it, too. Our DNA can increase our likelihood of getting it, but our lifestyle is still the main reason we do – or don't.

After discovering the Glucose Goddess Instagram account and learning about glucose and diabetes, Gustavo was ready for change, but the main barrier to it was his social life: when he went out for dinner, he would go along with his party and eat a lot of starches and sugars. He wanted to change his habits, but the judgment of his friends was difficult to deal with. 'Why are you ordering a salad?' they asked. 'Are you on a *diet*?'

So he devised a trick: before going out for dinner, at home, he made himself a big plate of grilled broccoli and ate it with salt and hot sauce.

With broccoli in his belly, he was ready for his meal out. When he got to a restaurant, he wasn't starving, so he could easily forgo the bread on the table. And anyway, the effect of whatever starch and sugar he now ate would be curbed by the broccoli. That meant less of a glucose spike and less insulin release, along with less inflammation, less damage to his cells and less inching towards type 2 diabetes.

Eighteen months into his glucose journey, Gustavo has lost 88 pounds. You'll learn about the other hacks he implemented in later chapters. When we spoke on the phone, he happily told me that he feels younger than ever. He can now run five miles without pain, something he had never been able to do but had always dreamed of. Beyond his physical improvements, Gustavo also shared that he felt more confident and better informed than ever: he explained that finally, he understands that *calories aren't everything*.

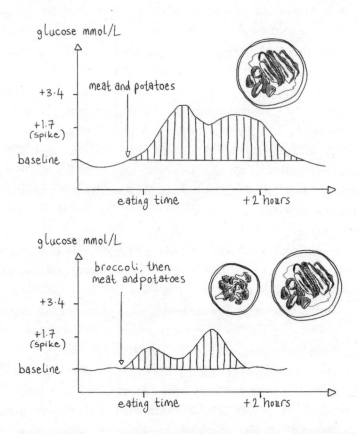

If you aren't sure you're going to be able to have a green starter at a restaurant, you can eat one before going out. Gustavo eats a large head of broccoli at home before meeting his friends at the steak house.

# HACK 3: STOP COUNTING CALORIES

If you follow the hack from the previous chapter, you will start adding calories to your meal in the form of a green starter. If you're hoping to shed pounds, you may wonder: *Is that really a good idea? Won't adding calories make me gain weight?* The short answer is no. The long answer involves understanding more about the *types* of calories we're eating – and setting things on fire.

To measure how many calories are in, say, a doughnut, here's what to do: dehydrate the doughnut and place it in a cubicle submerged in a water bath. Then set the doughnut on fire (yes, really) and measure by how many degrees the water around it warms up. Multiply the temperature change by the amount of water in the container, the energy capacity of water (which is 1 calorie per gram per degree), and you'll get the number of calories in the doughnut.

So when we say, 'This doughnut and this Greek yoghurt have the same number of calories', we're really saying, 'This doughnut and this Greek yoghurt warm up water by the same number of degrees when we burn them.'

It's through this burning technique – called a *calorimeter* and first invented in 1780 – that scientists can measure the calories in anything. The coal that your grandfather shovels onto the fire stands proudly at 3.5 million calories per pound (because it burns very slowly and releases a lot of heat). A 500-page book, on the other hand, isn't the best choice if you're trying to heat water: it contains only half a calorie (because a book turns to ashes very quickly and in the process doesn't produce much heat).

In any case, calories measure heat generated, nothing else.

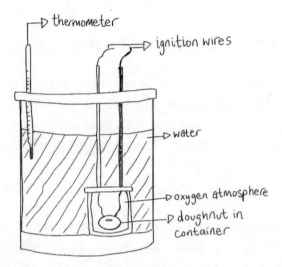

**To measure the calories in a doughnut, we measure how much it warms water up when burned.**

Judging a food based on its calorie content is like judging a book by its page count. The fact that a book is 500 pages long can certainly give you some information about how long it will take to read (about 17 hours), but it's unfortunately reductive. If you walk into a bookstore and tell an employee you want to buy 'a 500-page book', they will look at you a little strangely and then ask for clarifications. One 500-page book is not the same as another 500-page book, and likewise one calorie isn't the same as another calorie.

One hundred calories of fructose, 100 calories of glucose, 100 calories of protein and 100 calories of fat may release the same amount of heat when they burn, but they have vastly different effects on your body. Why? Because they are different *molecules*.

Here's this fact in action: in 2015, a research team out of UC San Francisco proved that we can keep eating the exact same number of calories, but if we change the *molecules* we eat, we can heal our body of disease. They demonstrated, for example, that calories from fructose are worse than calories from glucose (it's because fructose, as you know from Part I, inflames our bodies, ages our cells and turns to fat more than glucose does).

This study involved obese teenagers. They were asked to replace the calories in their diet that came from fructose with calories from glucose (they replaced fructose-containing foods such as doughnuts with foods containing glucose but no fructose such as bagels). The number of calories they consumed was kept constant. What happened? Their health improved: their blood pressure improved and their triglycerides-to-HDL ratio (a key marker of heart disease, as we learned in Part II) improved. They started reversing the progression of their fatty liver disease and their type 2 diabetes. And this profound change in their health happened in *just nine days.*

The results were conclusive: 100 calories of fructose are worse for us than 100 calories of glucose. This is why it's always better to eat something starchy than something sweet – more on that in Hack 9, 'If you have to snack, go savoury'. If the study had cut fructose and replaced it with protein, fat and fibre (if the participants had replaced the doughnuts with Greek yoghurt and grilled broccoli, for example), you can imagine that the effects would have been even more positive.

So if you've ever heard that to get healthy you just need to cut calories, now you know that's not true. You can do much to heal your body by changing the molecules you eat but keeping the calories the same.

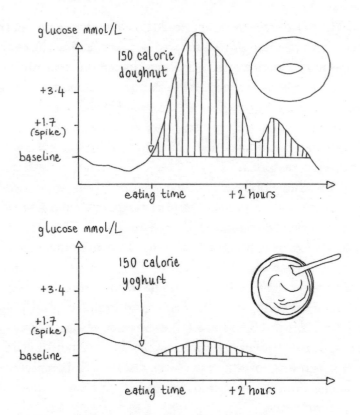

**Same calories, different effects. The calories from the doughnut (containing fructose) are preferentially converted to fat, inflame the body and age the cells. The calories from the yoghurt (no fructose) do so much less.**

How about weight loss – is it just a matter of consuming fewer calories? We used to think so, but that myth has been debunked, too. And there is a clue in the study I mentioned above: several of the teenagers in the study started losing weight even though they were eating the same number of calories as before. Impossible? No – but it definitely goes against what we have been told for years.

In fact, recent science shows that people who focus on flattening their glucose curves can eat *more* calories and lose *more* fat *more easily* than people who eat fewer calories but do not flatten their glucose curves. Let's repeat that: people on a glucose-flattening diet can lose more weight *while eating more calories* than people who eat fewer calories but spike their glucose levels.

For instance, a 2017 study from the University of Michigan showed that when overweight subjects focused on flattening their glucose curves (even if they ate more calories than the other group did), they lost more weight (17 pounds versus 4 pounds) than the subjects who ate fewer calories and took no care with their blood sugar levels.

It has to do with insulin: when we decrease our glucose levels, our insulin levels come down, too. A 2021 review analysing 60 weight loss studies proved that insulin reduction is primordial and always precedes weight loss.

In fact, it seems that we can completely ignore calories and still lose weight if we just focus on flattening our glucose curves. Keep in mind that this needs to be done with a little bit of good judgment (if you eat 10,000 calories of butter a day, your glucose curves will be flat, but you'll gain weight). The feedback on this from Glucose Goddess community members has been pretty much universal: if they take care not to spike their glucose levels, they can eat until they feel full, without counting calories, and still lose weight.

That was exactly what Marie did, and it changed her life.


## Meet Marie: she can't stop snacking

Twenty-eight-year-old Marie lives in Pittsburgh and works in operations at a tech company. For almost a decade, every time she left the house, she tucked a bag full of snacks under her arm. That was non-negotiable: if she didn't eat every 90 minutes, she would start feeling shaky and spacy and need to sit down. Her diary was organised around that requirement – if an event lasted longer than an hour and a half and she knew she wouldn't be able to eat during it, she would not attend. (She made an exception for her niece's baptism – but ate a cereal bar right before she entered the church and ran to her car at the end to open a bag of crisps.)

Many of us know someone (or indeed, *are* someone) who doesn't feel well if they don't eat at very specific intervals. People who experience this will sometimes say, 'I have low blood sugar.'

This is not necessarily incorrect – but what they might not know is that this isn't a condition that they were born with. More often than not, their low blood sugar is caused by the insulin released after a previous snack. What would be more accurate would be to say, 'My glucose levels are crashing.'

Usually, when insulin ushers glucose to the 'storage units' after a spike, the curve is smooth and bell-shaped, and glucose is brought back down steadily to its fasting level.

Sometimes, however, our pancreas releases too much insulin. As a result, too much glucose is stashed away. Instead of our glucose coming back down to fasting levels, it actually crashes and goes below normal for a while.

This is called *reactive hypoglycaemia*. When our glucose level dips and before our body brings it back up by releasing extra glucose into the blood, we can feel side effects: hunger, cravings, shakiness, light-headedness, or tingling in our hands and feet. Which was how Marie felt many times throughout the day.

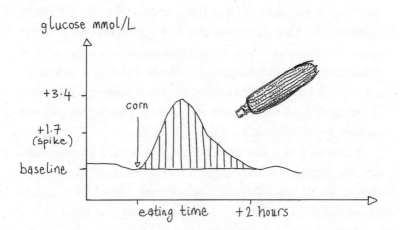

glucose mmol/L

+3.4

+1.7
(spike)

baseline

corn

eating time      +2 hours

**This is an example of insulin bringing glucose levels back down to normal after eating. After the spike, glucose went back down to its baseline level.**

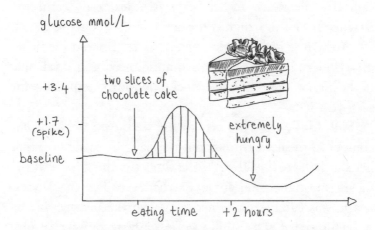

glucose mmol/L

+3.4

+1.7
(spike)

baseline

two slices of
chocolate cake

extremely
hungry

eating time      +2 hours

**This is an example of reactive hypoglycaemia and the hunger produced by it. After the spike, glucose levels crashed down way below the baseline level.**

Reactive hypoglycaemia is a common condition, especially in people with other glucose-related issues such as polycystic ovary syndrome (PCOS). How little or how much you experience it varies widely. In people with diabetes, the swings of reactive hypoglycaemia tend to be more pronounced – and their glucose level can get so low that it causes a coma. In people without diabetes, just a small dip can lead to extreme hunger, even if a meal was only two hours ago. And the greater the dip, the more hungry we become before the next meal.

A doctor's test confirmed that Marie did indeed have reactive hypoglycaemia. (The test involves drinking a shake with a lot of glucose in it and getting your blood values tested for three hours afterwards to detect the drop below baseline.)

This diagnosis was added to the long list of health conditions that Marie had developed since her teenage years: hypothyroidism, psoriatic arthritis, oestrogen dominance, candida infections, rashes, psoriasis, leaky gut, chronic fatigue, insomnia, night-time anxiety. On one occasion, when she had gone to pick up her latest thyroid medicine prescription, her pharmacist had announced that it was one of the highest doses he had ever compounded – especially for a 28-year-old.

Still, Marie tried her best to feel well. And since she felt compelled to snack throughout the day, she made sure that her snacks were 'healthy'. At the time, she thought 'healthy' meant mostly vegetarian foods that were low in calories. Marie was careful about her calorie intake in general (she never exceeded the 2,000-calorie daily recommendation), and she also forced herself to walk 10,000 steps every morning.

Her typical day went something like this: fruit and granola immediately upon waking at 5a.m. (she woke up that early because she was so hungry). Low-fat fruit yoghurt at 6a.m. A 100-calorie pack of cereal at 8a.m. A Pop-Tart at 9.30a.m. A vegetarian wrap at 11a.m. A vegetarian sandwich for lunch with coconut water and a 100-calorie pack of pret-zels, then a 100-calorie pack of cookies 90 minutes later. At 4p.m. she ate a full pound of grapes – that's about 180 grapes. Crackers an hour before dinner, followed by lots of rice and some beans for dinner, then a piece of chocolate before bed. She was eating the 'right' number of calories, but was hungry all the time. She was chronically fatigued and couldn't muster the energy to do anything past noon each day. She was so tired that she drank *ten cups of coffee a day*.

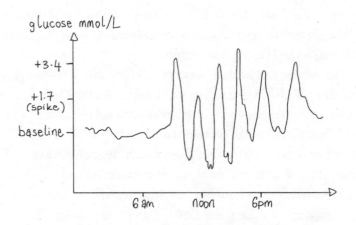

**This graph represents the glucose curve of someone with a diagnosis like Marie's – lots of spikes and dips below normal levels, called reactive hypoglycaemia.**

When people get a reactive hypoglycaemia diagnosis, they often hear that they should snack every few hours to make sure their glucose doesn't drop too low. But this just makes the problem worse: they eat something sweet or starchy, which shoots their glucose back up, releases insulin and makes their levels come crashing down again. Then the cycle repeats. They're on a never-ending glucose rollercoaster.

A more effective way to combat reactive hypoglycaemia

(which is a reversible condition, by the way) actually addresses the root problem: too much insulin. The solution is – you guessed it – to flatten the patient's glucose curve. With smaller spikes, that patient releases less insulin and experiences smaller dips. The body learns to not expect starchy and sweet snacks every few hours and, with less insulin present, starts burning fat reserves for fuel. It's important to make this change away from starchy and sweet snacks gradually, because it can take a few days or even weeks for the body to adapt.

That was what Marie desperately needed to do to feel better. Thankfully, while researching the meaning of glycaemia and going down an internet rabbit hole, she found her way to my Instagram account.

So Marie made some changes. Her plan was to eat as much as she felt was necessary as long as she kept her glucose curves flat. She ate her carbs last, added salads to her meals and introduced more protein, fat and fibre into her diet. She switched from mostly processed foods, made of sugars and starches and devoid of fibre, to mostly whole foods, with lots of fibre. She wasn't counting calories any more, but she definitely ate more than the 2,000 calories she used to.

Now, for breakfast, she has oatmeal with ground flax seeds, hemp seeds, nuts, pea protein powder and a sausage on the side. At lunchtime, two hard-boiled eggs, carrot sticks, celery, peanut butter or avocado, a protein smoothie (with collagen powder, 1 tablespoon of chia seeds, half a tablespoon of coconut oil and a whole bunch of greens), and half a banana last. For an afternoon snack, Greek yoghurt, berries and half a protein bar. Finally, at dinner, fish or chicken, kale sautéed with avocado oil and roasted sweet potatoes.

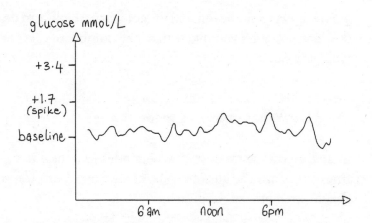

**This is what Marie's daily glucose levels look like now: small
variations within the optimal range. She eats more calories
than before, and she feels so much better.**

Marie shared the good news on the phone with me: 'I've
been able to go *four hours* without a meal! I can even exercise
fasted. That has given me my life back!'

That starving sensation every few hours quickly became a
thing of the past. So did her reactive hypoglycaemia. Other
things changed, too. Marie's energy levels increased within
a week or two, to the point where she went from ten cups of
coffee a day to one.

Her skin breakouts cleared up; her rashes and psoriasis,
too. Her headaches disappeared. So did her insomnia,
panic attacks and rheumatoid arthritis. Her oestrogen levels
returned to normal. She lost about five pounds. Her thyroid
functioning improved as well. Every couple of months, her
doctor administered tests and adjusted her meds to lower
and lower dosages. Her pharmacist doesn't remark on her
prescription any more. And almost best of all? She no longer
carries snacks in her bag. She doesn't need to. That may seem
like a small thing, but to Marie, it changed everything.

So remember this: health and weight loss have more to do with what molecules you ingest than the number of calories in what you eat.

## What does this mean for us?

It means we can, without fear, add calories to a meal if the calories help curb the glucose spike of that meal: i.e., if the molecules are fibre, fat, or protein.

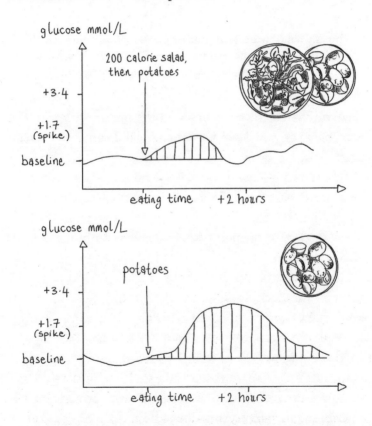

**When we add a 200-calorie salad (fibre and fat) to a meal, we are adding calories to it, but those calories help curb the glucose and insulin spikes. Those are good calories to add.**

When we add a dressed salad to a meal, the extra calories are useful to us, because they help keep our glucose and insulin levels low and even help us absorb fewer calories from what we eat after the salad (due to the mesh that fibre creates). On balance, we stay full longer, can burn more fat and put on less weight.

Flip the logic around: if we add more glucose or fructose to a meal, this *increases* the spikes, which leads to more weight gain, more inflammation and less satiety.

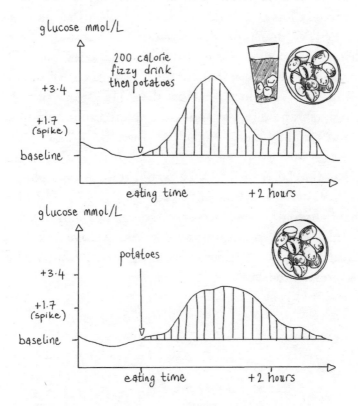

**When we add a 200-calorie fizzy drink (glucose and fructose) to a meal, those calories amplify the spikes: in fact, they increase the concentrations of the big three – glucose, fructose and insulin. Those aren't good calories to add.**

The fact that all calories are not equal is something that the processed food industry does its best to obfuscate. It hides behind calorie counts because they divert our attention away from scrutinising what's actually *in the box* – such as lots of fructose, which, unlike glucose, cannot be burned by our muscles for fuel and is almost all converted post-digestion into fat. Look at the 'shout lines' on snack packets when you next go to the store, and you will see what I mean.

This is exactly how Special K became a commercial success and to be perceived by consumers as the quintessential weight loss cereal: the box proudly advertised 'Just 114 calories!' We didn't think twice about it, and we didn't know that, even though it was relatively low in calories, Special K contained twice the amount of sugar as other cereals such as cornflakes. We didn't know that those 114 calories of sugar and starch would lead to a glucose spike and an insulin spike – and certainly more weight gain than 114 calories from, say, eggs and toast. We didn't know that those 114 calories of Special K for breakfast would start us on a glucose rollercoaster – and lead to cravings all day long. But now, thanks to continuous glucose monitors and curious scientists – which I'll tell you more about soon – we have proof that cereal for breakfast is definitely, unequivocally not a good way to start the day.

# HACK 4: FLATTEN YOUR BREAKFAST CURVE

The campus of Stanford University in California is home to a team of scientists who specialise in the study of continuous glucose monitoring. In 2018, they did something that all great scientists do: they challenged assumptions. Specifically, they set out to test the commonly accepted belief that, unless you have diabetes, your glucose levels should be of no concern. Second, and perhaps more controversially, they wanted to test a practice that has become a cultural norm: that cereal for breakfast is good for you.

Twenty participants were recruited, both men and women. None of them had been diagnosed with type 2 diabetes: their fasting glucose (as measured once a year by their doctor) was in the normal range. They arrived on a weekday morning at the lab to take part in the experiment – which consisted of eating a bowl of cornflakes with milk while wearing a continuous glucose monitor.

The results of this study were alarming. In those healthy individuals, a bowl of cereal sent their glucose levels into a zone of deregulation thought to be attainable only by people with diabetes. Sixteen of the 20 participants experienced a glucose spike above 7.8 mmol/L (the cut-off for prediabetes, signalling problems with glucose regulation), and some even spiked above 11.0 mmol/L (in the range of type 2 diabetes). That didn't mean that the participants had diabetes – they didn't. But it did mean that people without diabetes could spike as high as those with diabetes and suffer the harmful side effects those spikes cause. The discovery was groundbreaking.

The fact that a bowl of cereal causes spikes makes empirical sense. Cereal is made of either refined corn or refined wheat kernels, superheated, then rolled flat or puffed into various shapes. It's pure starch, with no fibre left. And because starch is not the most palatable thing on its own, table sugar (sucrose, made of glucose and fructose) is added to the concoction. Vitamins and minerals join the mix, but the benefit of these doesn't outweigh any of the harm of the other components.

In all, 2.7 billion boxes of cereal are sold every year in the United States alone. The most popular brand is Honey Nut Cheerios, which contains three times as much sugar as the cereal used in the Stanford study. So the alarming results those researchers observed are probably conservative compared to the glucose spikes taking place in the population at large.

When 60 million Americans eat a cereal such as Honey Nut Cheerios for breakfast in the morning, they are pushing their glucose, fructose and insulin levels into damaging ranges. Thus every day sixty million Americans are generating swarms of free radicals in their bodies, taxing their pancreas, inflaming their cells, increasing their fat storage and setting themselves up for a day full of cravings from shortly after they get out of bed.

Honestly, it's not their fault. Cereal is cheap, tasty and easy to reach for while still half asleep. My mom did it every day for a long time. Cereal looks harmless, but it's not. That goes for muesli, too.

Because of the way we eat today, early-morning spikes seem to be the norm. Whether it's cereal, toast and jam, croissants, granola, pastries, sweet oats, biscuits, fruit juice,

Pop-Tarts, fruit smoothies, acai bowls, or banana bread, the typical breakfast in Western countries is composed of mostly sugar and starch – a ton of glucose and fructose.

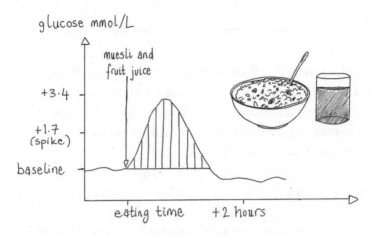

**In the United States, the typical breakfast is a bowl of cereal and fruit juice. Big spike.**

It's a common assumption that eating something sweet for breakfast is a good thing because it will *give us energy*. That's what I thought, growing up, when I spread Nutella on a crepe each morning. But that's actually not correct: though eating something sweet will give us *pleasure*, it's not the best way to give us *energy*.

Why? Well, as you know, when we eat glucose, we trigger insulin production. Insulin wants to protect us from the onslaught of glucose, so it removes it from circulation. So instead of the newly digested molecules staying around in our system to be used for fuel, they are stored away – as glycogen or fat.

Scientific experiments confirm this: if you compare two diets, the one with more carbohydrates leads to less available circulating energy post-digestion.

And that's not all I'm going to debunk here. You know the saying 'Breakfast is the most important meal of the day'? It's true, but not in the way you might think.

## How your breakfast secretly controls you

If we hit our foot sharply on the corner of our dresser while dancing around our bedroom, we feel it. It hurts. (I once broke a toe doing that.) We ice it, wrap it, yet still, it may well swell up so much that we can't wear our usual shoes. That might put us in a bad mood.

If a colleague or a family member asks us, 'What's wrong?' we can explain it plainly: I hurt my toe this morning, and that's why I'm grumpy. The connection is clear.

When it comes to how food affects us, the connection is murkier. We don't instantly *feel* the hurt that a spiky breakfast has on us. If, as soon as we ate that bowl of cereal, we were to have a panic attack, then fall asleep at the table, we'd get it. But because metabolic processes take hours to unfold, compound over time and become mixed with all the other things that happen in a day, connecting the dots takes a bit of detective work – at least until we get the hang of it.

A breakfast that creates a big glucose spike will make us hungry again sooner. What's more, that breakfast will deregulate our glucose levels for the rest of the day, so our lunch and dinner will also create big spikes. This is why a spiky breakfast is a one-way ticket to the glucose rollercoaster. A flat breakfast, on the other hand, will make our lunch and dinner steadier.

On top of that, first thing in the morning, when we are in our fasted state, our bodies are the most sensitive to glucose. Our sink (or stomach) is empty, so anything that lands in it will be digested extremely quickly. That's why eating sugars and starches at breakfast often leads to the biggest spike of the day.

Breakfast is the *worst time* to eat just sugar and starches, yet it's the time at which most of us eat *just* sugar and starches. (It's much better to have sugar as a dessert *after* a meal – I'll tell you more in Hack 6, 'Pick dessert over a sweet snack'.)

---

TRY THIS: Write down your typical breakfast ingredients. Which ones are starches? Which are sugars? Are you eating just sugars and starches for breakfast?

| I usually eat ... | Sugars | Starches | Protein, fat, or fibre |
|---|---|---|---|
| Example: orange juice | ✓ | | |
| Example: oats | | ✓ | |
| Example: butter | | | ✓ |
| | | | |
| | | | |
| | | | |

---

In speaking with people who've changed their diet to keep their glucose levels steadier, I have learned that this breakfast hack is key. Choose your breakfast well, and you'll feel better throughout the day – more energy, curbed cravings, better mood, clearer skin and on and on. It took a little while for Olivia to discover this, but once she did, there was no turning back.

## Good sugar, bad sugar and Olivia

Symptoms of dysregulated glucose can affect us at any age.

Olivia, aged 18, from a village near Buenos Aires, Argentina, was already experiencing an assortment of them: cravings for sweets (such as *dulce de leche*), bad acne on her forehead, anxiety, feeling drained in the evening but unable to fall asleep.

Olivia had gone vegetarian two years before, at 16, to reduce her carbon footprint. Unfortunately, as I explained earlier, the fact that a plate of food is vegetarian (or vegan or gluten-free or organic) does not mean that it's good for you. All of us, regardless of our diet, should also think about our glucose levels.

When she talked about her symptoms with friends, they told her she should eat something healthier in the morning, something with vitamins in it. They suggested a fruit smoothie to replace her usual toast spread with jam and hot chocolate. They explained to her that there was 'bad sugar' from chocolate and 'good sugar' from fruit. Olivia listened. Soon she started each morning with a fruit smoothie she blitzed at home – banana, apple, mango, kiwi.

Many people believe that some sources of sugar (namely fruits) are good for us and it's only the refined sugars in sweets, cakes and confectionary that are the bad for us.

Indeed, we've been indoctrinated with the idea. A century ago, the California Fruit Growers Exchange (which represents orange producers in the United States), which later became Sunkist, created a national campaign promoting the consumption of a daily dose of orange juice for its 'health giving vitamins and rare salts and acids'. But it forgot to mention that fruit juice is very bad for us and we can get vitamins and antioxidants in dozens of other foods that don't hurt us in the process. Unfortunately for Olivia, her friends had fallen for the same story. They thought that anything made from fruit was a healthy option.

To think that is to misunderstand the nature of sugar

– because *sugar is sugar*; it's the same whether it comes from corn or beets and has been crystallised into white powder, which is how table sugar is made, or from oranges and kept in liquid form, which is how fruit juice is made. Regardless of which plant they come from, glucose and fructose molecules have the same effect on us. And denying that fruit juice is harmful because of the vitamins it contains is a game of dangerous deflection.

What is true, however, is that if we are going to eat some sugar, a *whole* piece of fruit is the best vehicle for it. First, in a whole piece of fruit, sugar is found in small quantities. And you'd be hard pressed to eat three apples or three bananas in one sitting – which is how much can be found in a smoothie. Even if you did eat three apples or three bananas, you would take some time to eat them, much longer than it would take you to drink them in a smoothie. So the glucose and fructose would be digested much more slowly. Eating takes longer than drinking. Second, in a whole piece of fruit, sugar is always accompanied by fibre. As I explained earlier, fibre significantly reduces the spike caused by any sugar we eat.

By blending a piece of fruit, we pulverise the fibre into tiny particles that can't fulfil their protective duties any more. In case you're wondering, this does not happen when we chew – our jaws are powerful, but not as powerful as a blender's metal blades making 400 rotations *per second*. As soon as we blend, squeeze, dry and concentrate the sugar and remove the fibre in fruit, it hits our system fast and hard – and leads to a spike.

The more denatured a fruit is, the worse it is for us. An apple is better for us than apple sauce, which is better for us than apple juice. Essentially, as soon as it's juiced, dried, candied, canned, or turned to jam, you should think of fruit as *dessert*, just as you would a slice of cake. One 300ml bottle of orange juice (whether freshly squeezed, bought, with or without pulp) contains around 25 grams of sugar – that's the concentrated sugar of three whole oranges, with none of the fibre. It's the same amount of sugar as in a can of Coca-Cola. With 300ml of orange juice, you've reached the limit of the number of grams of sugar you should consume in a day, according to the American Heart Association (it recommends no more than 25 grams for women and 36 for men).

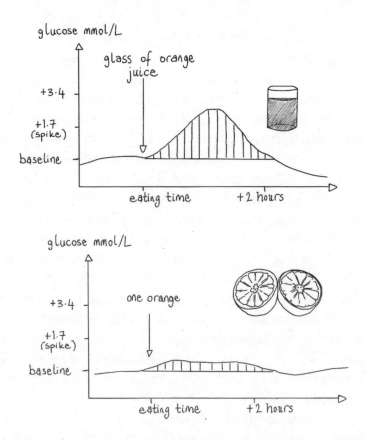

Yes, fruit juice has vitamins in it, but that's as much of a reason to drink it as the antioxidants in wine are a reason to drink alcohol.

Little wonder that with her new breakfast, things did not get better for Olivia. But she kept drinking smoothies day after day. The result? Worse acne, less energy and an even harder time falling asleep at night. Why did she feel as though she was getting worse when she was trying harder than ever to do things right? Because her smoothie was actually creating a bigger spike than her previous breakfast had.

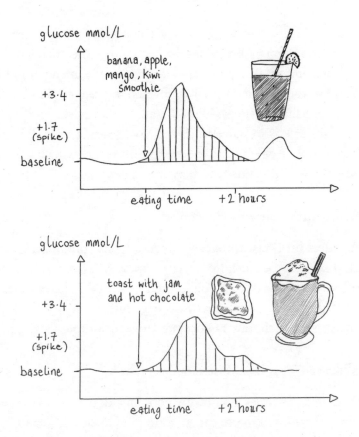

**Most of us think that a fruit smoothie is healthier than a breakfast with a mug of hot chocolate. In reality, when fruits are processed, they are no better than chocolate. Smoothies can be done right if they include other ingredients along with fruit. More on the ideal smoothie recipe on page 161.**

Olivia found the Glucose Goddess Instagram account. She recognised that she was feeling the symptoms of glucose spikes. And it was with great relief that she learned that what she thought was a smart choice – her fruit smoothie – was actually not. What did she do? She went savoury.

## Go savoury

The best thing you can do to flatten your glucose curves is to eat a savoury breakfast. In fact, most countries have a savoury option: in Japan, salad is often on the menu, and in Turkey, you'll find meat, veggies and cheese; in Scotland, smoked fish; and in the United States, omelettes.

This hack is so powerful that if you go savoury for breakfast, you'll be able to eat sweet later in the day with few side effects – and I'll show you how in the next hacks.

### Build your savoury breakfast plate

An ideal breakfast for steady glucose levels contains a good amount of protein, fibre, fat and optional starch and fruit (ideally, eaten last). If you're buying breakfast at a coffee shop, get avocado on toast, an egg muffin, or a ham and cheese sandwich, not a chocolate croissant or toast and jam.

### Make sure your breakfast contains protein

And no, this does not mean gobbling down 10 raw eggs every morning. Protein can be found in Greek yoghurt, tofu, meat, cold cuts, fish, cheese, cream cheese, protein powder, nuts, nut butter, seeds, and, yes, eggs (scrambled, fried, poached, or boiled).

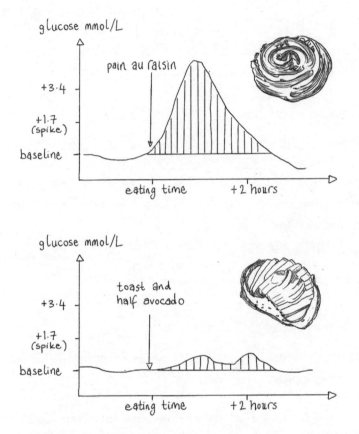

glucose mmol/L

pain au raisin

+3.4

+1.7
(spike)

baseline

eating time        +2 hours

glucose mmol/L

toast and
half avocado

+3.4

+1.7
(spike)

baseline

eating time        +2 hours

To make a healthy breakfast, just go savoury. Two breakfasts
with the same number of calories have vastly different
effects on glucose (and therefore insulin) levels. In the top
graph, a breakfast of starch and sugar leads to weight gain,
inflammation and return of hunger shortly thereafter.
A breakfast of starch and fat (bottom graph) has none
of these side effects.

## Add fat

Scramble your eggs in butter or olive oil, add slices of avocado, or add five almonds, chia seeds, or flaxseeds to your Greek yoghurt. By the way, never eat fat-free yoghurt – it won't keep you full, and I'll explain why later. Switch to 5 per cent regular or Greek yoghurt.

## Extra points for fibre

It can be challenging to get fibre in the morning because it means eating veggies for breakfast. I don't blame you if you aren't into that. But if you can, try. I love mixing spinach into my scrambled eggs or tucking it underneath a sliced avocado on toast.

Literally any vegetable will do, from spinach, mushrooms and tomatoes to courgettes, artichokes, sauerkraut, lentils or lettuce.

## Add starch or whole fruit for taste (optional)

This can be oats, toast, rice, or potatoes or any whole fruit (the best option is berries).

Olivia decided to try the savoury breakfast hack for herself. First thing the next day, she bought some eggs. To get some ideas of what else to put on her plate, she thought about her favourite lunch and dinner ingredients, and the result was a tasty dish: an omelette with avocado, sunflower seeds, olive oil and a dash of sea salt. Very soon, she felt the difference in her body – she felt lighter, less bloated, healthier and full of energy.

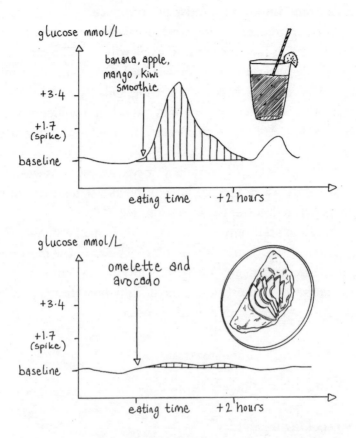

**The tradition that breakfast should be sweet is completely misguided. Build your breakfast around protein, fat and fibre for satiety and stable energy.**

It wasn't just her body. It was her brain, too. Her studying (she's in second year of design school) got sharper, and her grades got better.

Scientists have tried to measure how various breakfasts impact our performance on cognitive tests. And the answer to whether sugar makes your brain work better is... no. A review of 38 studies couldn't draw any definitive conclusions but stated that, if anything, a breakfast with a flatter curve could improve cognitive performance.

Furthermore, the curve created by your first meal of the day influences how the rest of the day will go. With no spike, you'll ride into the afternoon with satiety and stable energy, as Olivia learned to do. With a big spike, you'll set off a chain reaction of cravings, hunger and lagging energy until the evening. And these chain reactions compound day after day.

So if you want to improve just one aspect of your daily food habits, eat a glucose-healthy breakfast for maximum impact. You'll notice the effects immediately.

Really, it's one of the most practical changes to make. You can plan ahead. Your willpower is at its best in the morning. And there are usually no friends around to tempt you to abandon it. I promise, a glucose-healthy breakfast can be as easy to put together as a bowl of cereal.

### The 5-minute savoury breakfast

You can mix and match any of these.

**No cooking involved:**

- A bagel with cream cheese, topped with a few lettuce leaves and slices of turkey
- A can of tuna, a few pecans and olives, a drizzle of olive oil
- An apple with walnuts and slices of Cheddar
- Full-fat yoghurt with sliced fruit such as a peach, a drizzle of tahini and salt
- Greek yoghurt swirled with 2 tablespoons of nut butter and a handful of berries
- Half an avocado with 3 tablespoons of hummus, lemon juice, olive oil and salt

- Homemade granola that is nut-centric or cereal designed specifically with extra fibre or protein (see the Cheat sheets after Hack 10 to learn how to decipher packaging)
- Slices of ham on crackers
- Slices of smoked salmon, avocado and tomato
- Toast with almond butter
- Toast with mashed avocado
- Tomato and mozzarella with a drizzle of olive oil
- My go-to: leftovers from last night's dinner – the fastest option of them all!

**Cooking involved:**

- A tortilla filled with black beans and chopped avocado
- A full English breakfast (eggs, sausage, bacon, beans, tomatoes, mushrooms, toast)
- Hardboiled eggs with hot sauce and avocado
- Pan-fried halloumi cheese, tomatoes, salad
- Poached eggs with a side of sautéed greens
- Quinoa porridge topped with a fried egg
- Sausage and grilled tomatoes
- Scrambled eggs with crumbled goat's cheese
- Toast topped with a fried egg
- Warm lentils topped with a fried egg

## The still-sweet breakfast

If you're not ready to say goodbye to a sweet breakfast (or if you're staying with a particularly pushy aunt who likes to make pancakes from scratch in the morning), here's what to do: eat the sweet things *after* something savoury.

First, eat protein, fats and fibre – an egg, for example, a couple of spoonfuls of full-fat yoghurt, or any combination of the foods from 'The 5-minute savoury breakfast' section above. *Then* have the sweet food: cereal, chocolate, French toast, granola, honey, jam, maple syrup, pastries, pancakes, sweetened coffee drinks. For example, if I really want some dark chocolate when I wake up (What? It happens.), I have it *after* a plate of eggs and spinach.

Remember the sink analogy of Hack 1, 'Eat foods in the right order'? With a stomach that contains other things, the impact of that chocolate or sugar and starch will be lessened.

## Sweet breakfast cheat sheet

Can't go without something sweet in the morning? Here are some ways to eat that something but reduce any spike it could cause.

### Oats

If you love oats (which are starch), eat them alongside nut butter, protein powder, yoghurt, seeds and berries. Avoid adding brown sugar, maple syrup, honey, tropical fruits or dried fruit.

You can also try a chia pudding instead: chia seeds soaked overnight in unsweetened coconut milk with a spoonful of coconut butter.

### Acai bowls

Acai bowls – traditionally a Brazilian dish but now eaten around the world – are essentially a thick berry smoothie

# GLUCOSE REVOLUTION RECIPES

Creamy steel-cut oats with nut butter, berries and cacao nibs (p249)

Green shakshuka of leek, fennel and spinach with tahini and dill (p250)

Green Goddess soufflé omelette with rocket, avocado and smoked salmon (p251)

Tenderstem broccoli with basil, lemon, chilli and Parmesan (p257)

Tangy red cabbage with pomegranate seeds and coriander (p255)

Simple green salad with artichoke, hazelnuts and feta (p254)

Chicken traybake with baby potatoes, olives, capers and cherry tomatoes (p258)

Cod with tahini, pine nuts and spinach (p261)

Greens and grains salad with chicken, pomegranate, pistachios and a lemon yoghurt dressing (p262)

Linguine with oven-baked feta, oregano and tomatoes (p266)

Spicy pork tacos with black bean, tomato and lime salsa, avocado and baby gem (p267)

Jessie's grandmother's pot-baked white cabbage with ribeye steak (p264)

Baked rhubarb posset with almonds (p271)

Gooey chocolate brownie with smashed raspberry Greek yoghurt (p276)

topped with granola, fruit, and other ingredients. They sound healthy because they are made from fruit, but now you know that that in no way makes something healthy. On a closer look you'll find that they are composed entirely of sugars and starches. So apply the same guidelines as for oats, above.

If you are wondering about agave and honey and how they compare to low-calorie sweeteners, we'll cover it next in Hack 5, 'Have any type of sugar you like – they're all the same'.

## Smoothies

You *can* enjoy a smoothie for breakfast; it just has to incorporate protein, fat and fibre. Start your smoothie with protein powder, then add a combination of linseed or flaxseed oil, coconut oil, avocado, seeds, nuts and a cup of spinach. Finally, add some sugar for taste: ideally berries, which add a sweet taste but are significantly higher in fibre than other fruits.

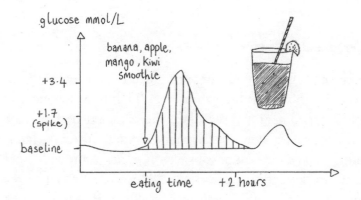

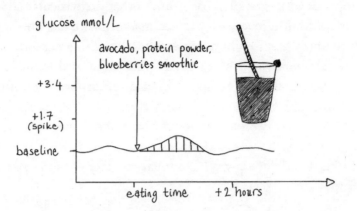

**The more protein, fat and fibre and the less fruit your smoothie contains, the better it will be for your glucose levels.**

A good rule of thumb for a smoothie: don't put more fruit into the blender than you could eat whole in one sitting. My go-to smoothie recipe is 2 scoops of protein powder, 1 tablespoon of flaxseed oil, ¼ avocado, 1 tablespoon of crunchy almond butter, ¼ banana, 1 cup of frozen berries and some unsweetened almond milk.

## Cereal and granola

Some cereals are better for your glucose levels than others. Look for those that brag about their high fibre content and low sugar content. (In the Cheat sheets included after Hack 10, I'll explain how to decipher nutrition labels on packaging to pick the best possible cereal.) Then eat it with 5 per cent Greek yoghurt instead of milk, which adds fat to the mix. Top with nuts, hemp seeds and/or chia seeds to add protein too. If you need to sweeten it, do so with berries – not sugar.

Granola may seem healthier, but it's usually just as full of sugar as other cereals. If you love it, look for a low-sugar granola with a high nut and seed content – or even better, make your own.

For cereal addicts: you can still have it in the morning if it's not the centre of your breakfast. Here's an idea: eat it last, after something that has protein in it.

### Fruit

The best options to keep your glucose levels steady are berries, citrus fruits and small, tart apples, because they contain the most fibre and smallest amount of sugar. The worst options – because they have the highest amount of sugar – are mangoes, pineapple and other tropical fruit. Make sure you eat something else before them.

### Coffee

Beware of sweetened coffee drinks – and know that cappuccinos are better for your glucose levels than mochas, which contain chocolate and sugar. If you like to have a sweetened coffee drink, try mixing the coffee with full-fat milk or cream (fat is not to be feared) and sprinkling cocoa powder on top instead. Non-dairy milks made from almonds or other nuts work, too; they are better than oat milk, which produces bigger spikes, as it's made from grains, which contain more carbs. If you add sugar to your coffee, make sure you eat something glucose-steady beforehand – even just a slice of cheese. And if you're wondering if certain sweeteners are preferable to others, read on.

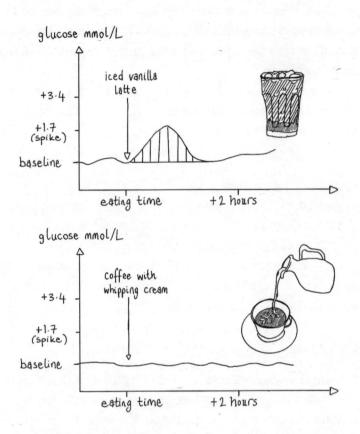

**Sweetened coffee drinks can drive a big spike. Opt for cappuccinos, americanos, macchiatos and unsweetened lattes over coffees with flavours, syrups – and sugar.**

### What if I don't eat breakfast?

No problem. The same concept applies – whenever the first meal of your day is, make it savoury to set yourself up for success.

### Should I try to eat my breakfast ingredients in the right order, too, as specified in Hack 1?

Ideally, yes, but don't stress if you can't. The hacks in this book should be used when it's easy. If it's a bowl of full-fat yoghurt topped with a seed- and nut-filled granola and you want to eat those components together, go for it. You're already making a good choice by choosing that over cereal.

### Aren't eggs bad for your heart?
Scientists used to think that eating foods that contain cholesterol (such as eggs) increased our risk of heart disease. Now we know that's not true – as we learned in part II, sugar is actually the bad guy. Research shows that when people with diabetes replace their oatmeal with eggs at breakfast (and keep their calories constant), their inflammation and risk of heart disease go down.

> TRY THIS: Treat your breakfast like your lunch, and eat a savoury meal. What happens? How do you feel?

## Let's recap

Eating cereal in the morning has become a habit for many of us, but as you've learned in these pages, a sweet breakfast is a ticket to a glucose rollercoaster. Eating a savoury one will help curb hunger, banish cravings, boost energy, sharpen mental clarity and more for the next 12 hours.

Cereal for breakfast is just one of the habits I'm here to debunk. The next one has to do with adding sugar, honey and sweeteners to our food and drinks – and the fact that the common assumption about which is 'healthiest' is wrong.

# HACK 5: HAVE ANY TYPE OF SUGAR YOU LIKE – THEY'RE ALL THE SAME

You know that famous line from *Romeo and Juliet*: 'A rose by any other name would smell as sweet'? Well, you could say the same thing about sugar. Sugar by any other name still has the same impact on our body.

## Is honey healthier than sugar?

As you learned in Hack 3, 'Stop counting calories', when it comes to understanding what a food does to our bodies, it is the molecules that matter, not the calories. There is something else that doesn't matter: the *name* of the food.

This is surprising to most people, but on a molecular level, there is no difference between table sugar and honey. And there is no difference between table sugar and agave syrup. In fact, there is no difference between table sugar and any of these: agave syrup, brown sugar, caster sugar, coconut sugar, icing sugar, demerara sugar, evaporated cane juice, honey, maple syrup, molasses, muscovado sugar, palm sugar, palmyra tree sugar. They are all made of glucose and fructose molecules. They are just packaged differently, named differently, and priced differently.

Honey begins as nectar from plants, but it contains glucose and fructose, just as table sugar does. Brown sugar (which sounds healthy, right?) is made of the exact same thing as white sugar, except that it is tinted (yes, *tinted*) with molasses, a by-product of the sugar-making process, to make it look more wholesome.

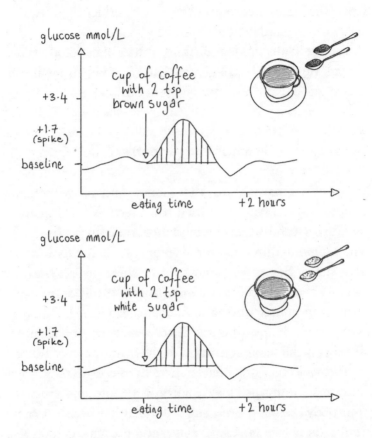

**Many of us believe that brown sugar is better for us than white sugar. There is actually no difference.**

Muscovado sugar is even darker than brown sugar because it has even more molasses in it. Caster sugar and icing sugar are table sugar ground into a finer powder. Demerara and cane sugar are golden in colour because they have been bleached less during the refinement process. Coconut sugar is sugar from a coconut instead of from a cane or a beet.

Palm (or palmyra tree) sugar is from a palm. The list goes on.

And misinformation is rampant: for instance, the Philippines, a large producer of coconut sugar, released data claiming that coconut sugar was healthier than regular sugar, which was later proven to be wrong.

You get it: any kind of sugar, regardless of its colour, taste, or plant of origin, is still glucose and fructose, and will still lead to glucose and fructose spikes in our bodies.

## Is natural sugar better?

Many of us have heard that honey and agave contain 'natural' sugars. And that dried fruit, such as dried mango, contains 'natural' sugars because they come from a fruit. It's, um, *natural* to think that those options are better for us than table sugar. But here's something to chew on: all sugar is natural, because it always comes from a plant. Some table sugar even comes from a *vegetable* (sugar beets). But that doesn't make it any different. There is no good or bad sugar; all sugar is the same, regardless of the plant it comes from.

The molecules are what matter: by the time they reach your small intestine, they're all just glucose and fructose. Your body doesn't process sugar differently whether it came from a sugar beet, an agave plant, or a mango. As soon as a fruit is denatured and processed and its fibre is extracted, it becomes sugar like any other sugar.

It is true that in dried fruit, there is still some fibre present. But because all the water is taken out of the fruit, we eat many, many more pieces of dried fruit than we would pieces of whole fruit. So we consume way more sugar way more quickly than nature intended, resulting in big glucose and fructose spikes.

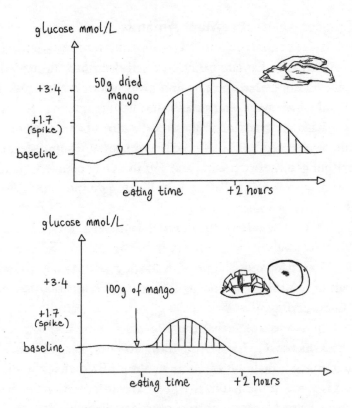

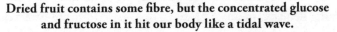

**Dried fruit contains some fibre, but the concentrated glucose and fructose in it hit our body like a tidal wave.**

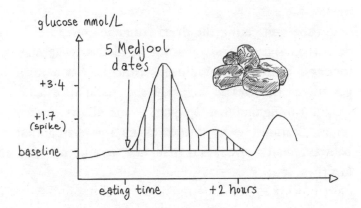

**Sugar is sugar. Choose whole fruits over dried fruit.**

# Meet Amanda

Amanda is in her late twenties, a self-described 'health nut', who watches what she eats and loves working out regularly – and continued doing so well into her first pregnancy. That was why her gestational diabetes diagnosis came as a shock. She was scared, for both herself and her baby – and she also felt judged by her friends and family. They couldn't believe her diagnosis, either. *What, you? We thought you were healthy! How's that possible?*

As the months progressed towards her due date, her glucose levels kept rising, and her insulin resistance got worse. She felt out of control. And she really thought she was eating healthily – including lots of dried fruit to satisfy her sugar cravings.

She wrote to me that the information she had found on the Glucose Goddess Instagram account had helped her take back a little bit of control. She realised that the diagnosis was not her fault. The posts and information she read there helped her see that gestational diabetes happens to many healthy people. She learned about things she could do to flatten her glucose curves and avoid going on medication.

So she quit eating the dried fruit she used to eat every day. Then she went savoury for breakfast, swapping oats for eggs. Those small modifications helped her manage her gestational diabetes so well that she maintained a healthy weight throughout her pregnancy and didn't have to go on medication. I was thrilled when she shared that her baby boy had arrived, and that they were both happy and healthy.

### What about agave syrup's 'lower glycaemic index'?

During her pregnancy Amanda was also told that agave syrup was better for her than sugar because it had a lower glycaemic index. What's that about? Let's dive in.

Although sugar is sugar, regardless of its source, what *is* true is that the ratio of glucose and fructose molecules is different from sugar to sugar. Some sugars contain more fructose, while others contain more glucose.

And while agave syrup may be often recommended to people with diabetes and women diagnosed with gestational diabetes it actually contains much more fructose than table sugar (90 per cent compared to 50 per cent). This means that the fructose spike is bigger.

Now, get this: recall from Part I that fructose is worse for us than glucose: it overwhelms our liver, turns to fat, precipitates insulin resistance, makes us gain more weight than glucose and doesn't make us feel as full. As a result, *since agave has more fructose than table sugar does, it is actually worse for our health than table sugar.*

Don't believe the hype.


### But what about the antioxidants in honey?

In the same way that we saw in Hack 4 that there is no logic to drinking fruit juice for the vitamins, there is no logic to eating honey *for the antioxidants*. Yes, honey contains antioxidants and fruit juice contains vitamins, but those don't outweigh the impact of the large amounts of glucose and fructose they contain.

And fun fact: there aren't that many antioxidants in honey anyway; you can find all the antioxidants contained in a teaspoon of honey in *half a blueberry*. That's right – half a blueberry!

## The good news: pick any sugar you like

We don't need to eat sugar to live (recall that our body doesn't need fructose, just glucose, and it can make that from within if we don't eat it), and we don't need to eat sugar to get energy (remember, sugar actually *dampens* our energy levels).

Since all sugars, regardless of their source, are eaten for pleasure, pick whichever one you like best – and enjoy it in moderation. If you prefer the taste of honey over table sugar, go for it. If you prefer baking with brown sugar, fine too.

## And as much as you can, choose fruit for your sweet fix

When we want something sweet, the best thing we can eat is whole fruit. Remember, that's the way nature intended us to consume glucose and fructose – in small quantities, not too concentrated, in combination with fibre.

Have sliced apple in your oatmeal instead of table sugar, berries in your yoghurt instead of honey.

Other ingenious additions to either oatmeal or yoghurt include cinnamon, cacao powder, cacao nibs, shredded unsweetened coconut, or unsweetened nut butter (I know it sounds strange, but nut butter on its own tastes sweet and makes for a dessert-worthy combo).

## Artificial sweeteners

That's 'natural' sugars. What about artificial sweeteners?

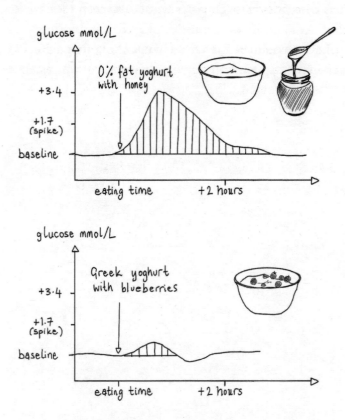

**A 5 per cent Greek yoghurt with blueberries is as sweet as, but so much better for your glucose curve than, a 0 per cent regular yoghurt with honey.**

Some artificial sweeteners spike our insulin levels, which mean they prime our body to store fat and encourage weight gain. For instance, research shows that when people switch from drinking diet drinks to drinking water, they lose more weight (two extra pounds in six months in one study) – without altering the number of calories they eat.

What's more, preliminary studies suggest that the taste of sweeteners may increase our craving for sweet foods, just as sugar does. The theory posits that we could then be more likely to satisfy those cravings because sweeteners have lower calorie counts, so we think it's okay to eat another cookie. Artificial sweeteners may also change the composition of our intestinal bacteria, with potentially negative consequences.

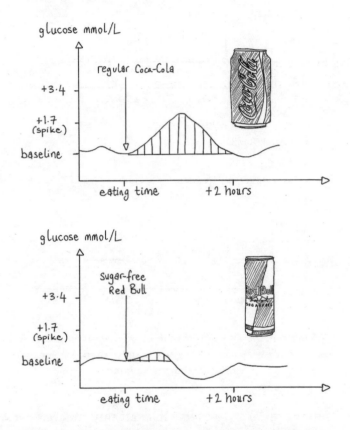

**Sugar-free Red Bull contains aspartame. Aspartame may create an insulin spike, though science doesn't yet have a definite answer. The aspartame could explain why I saw a dip in my glucose levels after drinking it – a surge in insulin leads to a drop in glucose.**

The best sweeteners that cause no side effects on glucose and insulin levels are:

- Allulose
- Monk fruit
- Stevia (look for pure stevia *extract* because some other forms of it are mixed with glucose-spiking fillers)
- Erythritol

There are some artificial sweeteners I'd recommend you avoid, because they are known to increase insulin and/or glucose levels, especially when combined with foods, or cause other health issues. They are:

- Aspartame
- Maltitol (turns to glucose when digested)
- Sucralose
- Xylitol
- Acesulfame-K

Sweeteners aren't a perfect replacement for sugar. Many people don't like their taste, and some even get a headache or stomach-ache from them. And really, they don't taste as good as sugar. Monk fruit in a breakfast shake is okay, but sometimes you just need the real stuff – when baking, for example. The best thing to do, in my opinion, is to use sweeteners to wean ourselves off the need to sweeten everything. Because sweetness is addictive.

### What about diet drinks?

Let's be clear: in a vacuum, it's better to drink artificially sweetened diet drinks than regular fizzy drinks. *But* diet drinks aren't the same as water. They contain artificial sweeteners, which can lead to some of the harmful outcomes I describe above.

## The addiction conundrum

It's easy to feel addicted to eating sweet things. I once felt addicted to sweet things, too. This feeling is not our fault – remember, sweetness activates the addiction centre in our brain. The more we eat it, the more we want it.

To slowly wean yourself off the taste, there are a few things you can do. Replace that spoonful of sugar in your coffee with allulose, then lower the quantity over time. Next time you want a sweet or piece of chocolate, try eating an apple. Or when a craving hits, notice it and take some deep breaths. In my experience, it usually passes after 20 minutes. But if you're still in the throes of it, try eating something else – usually something with fat, like cheese, does the trick. I also like drinking teas that are naturally sweet, such as cinnamon or liquorice. It always helps me.

And if you still want that sweet something, eating it without guilt is the best thing to do.

## Let's recap

It's very unlikely that we're going to get rid of sugar in our diets entirely. And I'm here to tell you that that's okay. A birthday isn't much fun if you serve Brussels sprouts instead of birthday cake.

What if, instead of labelling ourselves a failure every time we eat sugar, we were to be thoughtful about when we eat it and accept – joyfully – that it's part of our lives?

I eat sugar when my mom makes birthday cake (chocolate cake with a crackly, shiny, sugary crust), when my grand-mother makes *brigadeiro* (a delicious Brazilian dessert made from chocolate and sweet condensed milk), when I eat my favourite ice cream (Häagen-Dazs Belgian chocolate topped with two spoonfuls of chocolate fudge), or when I'm craving a piece of chocolate (can you guess I like chocolate by now?). The rest of the time, if I want something sweet, I eat berries, monk fruit, almond butter, or cacao nibs.

I'm often asked questions such as: 'I have honey and milk before I go to bed. Is that okay?' or 'Is it bad to add maple syrup when I eat pancakes?' To which I answer: eat it if you really love it and it's worth the corresponding glucose spike.

## Sugar is okay in moderation

We should also try to let go of the impossible-to-keep promises we make to ourselves. I've said things such as: 'Starting tomorrow, I'm never eating any cupcakes ever again.' Or: 'This is the last piece of chocolate I will buy.' But when we make foods off limits in an effort to force a lifestyle change, it doesn't work. The hour comes when we can't take it any longer and we empty the cookie jar.

We often think that if we can't do something perfectly, such as stick to a diet, we shouldn't do it at all. That couldn't be further from the truth. It's about doing your best, and as you'll start to feel better, your cravings will dissipate and you will be impressed by how easily your sugar intake will diminish.

I promised you in the previous chapter that I would show you how, if you skip sugar at breakfast, you can enjoy it later in the day. The next three hacks tell you how – in ways that will keep your glucose curves steady. That means you get to eat what you love without gaining as much weight, deepening your wrinkles, adding plaque to your arteries, or any of the other short- or long-term consequences of high glucose levels. It sounds like magic, but it's science.

# HACK 6: PICK DESSERT OVER A SWEET SNACK

After a meal, we tend to move on quickly to our next activity – maybe doing the dishes, getting back to work, or going about our day. But when we're done eating, our organs are just getting started – and they keep working for *four hours* on average after our last bite. This busy time is the 'post-eating,' or *postprandial*, state.

## What happens in the postprandial state

The postprandial state is the period of our day when the largest hormonal and inflammatory changes take place. To digest, sort and store the molecules from the food we just consumed, blood rushes to our digestive system and our hormones rise like a tide. Some systems can be put on hold (including your immune system), while others are acti-vated (such as fat storage). Insulin levels, oxidative stress and inflammation increase. The bigger a glucose or fructose spike after a meal, the more demanding the postprandial state is for our body to deal with, because it has more free radicals, glycation and insulin release to manage.

The postprandial state is normal, but it's also an effort for our body. Processing a meal can take more or less work, depending on the amount of glucose and fructose that we've just consumed. We tend to spend about 20 hours of a 24-hour day in the postprandial state, because we have on average three meals and two snacks a day. This used to be different: up until the 1980s, people didn't snack as often between meals, so they spent only eight to 12 hours in the postprandial state. Snacking is a 1990s invention, like low-waisted jeans (something to think about).

When our body is not in the postprandial state, things are a little easier. Our organs are on clean-up duty, replacing damaged cells with new ones and clearing our systems. For instance, the gurgling we feel in our small intestine when we haven't eaten in a few hours is our empty digestive tract cleaning its walls. When our body is not in the postprandial state, our insulin levels come down and we can go back to burning fat instead of stashing it.

You might have heard that back in prehistoric times, we could, if needed, go a long time without eating. It's because we could easily switch between using glucose for fuel (from our last meal) to using fat for fuel (from our fat storage). This switching ability, as mentioned earlier, is called *metabolic flexibility*. It is a prime measure of a healthy metabolism.

Remember Marie, who used to leave the house with a bag full of snacks? She was an example of low metabolic flexibility. She *needed* to eat every 90 minutes because her cells had come to rely on glucose for fuel every few hours. When Marie changed the way she ate, she retrained her cells to use fat for fuel instead. She could then go hours without eating. Marie increased her metabolic flexibility.

To increase your own metabolic flexibility, eat larger, more filling meals so you don't need to snack every hour or two. That goes against a popular belief that eating 'six small meals a day' is better than two or three big ones, but research bears it out. Scientists in the Czech Republic, in 2014, tested this in people with type 2 diabetes. They decided on a daily calorie quota and got one group of participants to consume their calories in two large meals and the other group to consume them in six small meals. The two-meal group not only lost more weight (8 pounds versus 5 pounds in three months) but also saw improvements in the key markers of their overall health: their fasting glucose decreased, their liver fat decreased, their insulin resistance decreased and their pancreas cells got healthier. Same calories, different effects. (Back to one of my favourite mantras: calories aren't everything.)

Another way to improve your metabolic health is via what's called *intermittent fasting*, where you either fast for 6, 9, 12, or 16 hours at a time or reduce your calorie intake a couple of days a week.

But this chapter is not about that. This chapter is about an insight from the latest research on glucose spikes: if you want to eat something sweet, it's better to have it as dessert rather than as a snack in the middle of the day on an empty stomach. Understanding the postprandial state is key to learning why.

## Why dessert wins

When we forgo snacks, we keep our system out of the post-prandial state for longer. That means there's time for the clean-up I described above. And by eating something sweet after a meal, we lessen the corresponding glucose spike it would cause because – shout-out to Hack 1 – eating sugar and starches *last*, after other food (instead of first, or on their own as a snack), means they move from sink to pipe more slowly.

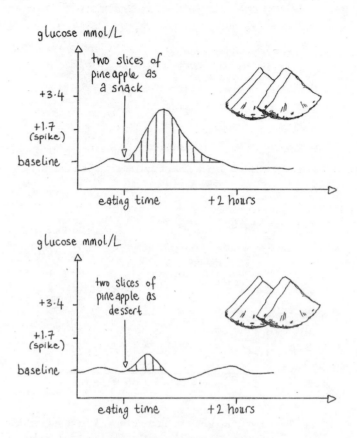

**Same pineapple, different spike. If it's eaten as a dessert after a meal containing fat, fibre and protein, a pineapple will create a smaller spike. We do see a small reactive hypoglycaemia, but this is less of an issue than the big spike when pineapple is eaten as a snack. Bigger spike, more symptoms.**

So whether it's a piece of fruit, a smoothie, a toffee, or a cookie, if you're going to eat it, do it at the end of a meal.

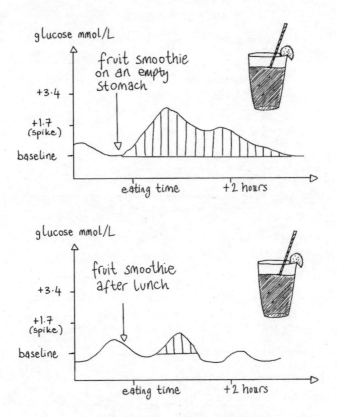

**It's all about reducing glucose fluctuations. A fruit smoothie on an empty stomach created a spike of about 2.8 mmol/L; after a meal, the overall variability it caused was lower.**

TRY THIS: If you feel the urge to eat something sweet between meals, put it aside – in the fridge or somewhere else – and enjoy it for dessert after your next meal instead.

# Meet Ghadeer

Ghadeer is a translator and mother of three who lives in Kuwait. She suffers from PCOS and has done so since her first period, at 13. She has contended with all its symptoms, from acne to mood swings to weight gain. She has experienced several miscarriages. A few years ago, at the age of 31, she was diagnosed with insulin resistance and her periods stopped altogether.

Her doctor encouraged her to change her lifestyle – to eat better and exercise more. But she was lost as to where to begin. It was pretty vague advice, as advice goes, and was received without much enthusiasm. Ghadeer didn't know what to do next, nor did she believe that what she did could manage her condition – until the day she came across the Glucose Goddess Instagram account.

There, it all clicked. Insulin resistance and PCOS are linked. Both have the same cause – dysregulated glucose levels. That information changed Ghadeer's life. On top of that, she was thrilled when she realised she could address her symptoms without going on yet another diet. She had been on what felt like a hundred diets – and she was tired of diets. She never wanted to go on one again.

So she tried a few of these hacks. She started to eat her food in the right order. She swapped fruit juice for tea. She replaced sugar for monk fruit. She didn't stop eating chocolate and sweets, which she loves, but she now eats them as dessert instead of as snacks. Her days now consist of three meals instead of three meals plus snacks.

In three months, her periods came back. Other changes: her average glucose level used to be 9 mmol/L; now it's 5 mmol/L. She lost over 20 pounds, and she has rid herself of PCOS symptoms and insulin resistance. She feels the difference in her mood, too: she's more patient with her kids. 'I have never, in my whole life, felt like this. I feel so good. My body is my friend now.'

The changes were so drastic that her doctor marvelled. 'What did you do?' he asked. She shared everything she had learned.

### Should I try to eat only once or twice a day?
You don't need to go that far. Some people find that this form of intermittent fasting suits them very well, others that it's hard to maintain. Studies have shown that the benefits are more pronounced for men and that for women of reproductive age, fasting too long and too often may cause hormonal disruptions and other types of biological stress. Try three meals a day, and see how you feel.

### What about late-night snacking?
If you typically have a sweet snack a few hours after dinner, a better alternative is to have it as dessert after you finish your main. If a late-night snack is inevitable, read on for other hacks that will help.

### How do I know if I'm metabolically flexible?
If you can easily go five hours between meals without feeling light-headed, shaky, or hangry, you're likely to be metabolically flexible.

## Let's recap

The best time to eat something sweet is after you've already eaten a meal with fat, protein and fibre. When we eat sugar on an empty stomach, we're throwing our system into a postprandial spin, riding on a big glucose and fructose spike. But if you can't avoid eating sugar on an empty stomach – a last-minute birthday party invite, a workday bake-off, an ice cream date with your crush – I'm here for you. Read on to discover another supercool hack.

# HACK 7: REACH FOR VINEGAR BEFORE YOU EAT

Do you want to drizzle some vinegar on your brownie? Didn't think so. Don't worry, that's not what I am about to suggest. I'm talking about mixing a vinegar drink and sipping it before you eat your next sweet treat – whether for dessert or on the occasions when you have it as a stand-alone snack.

The recipe is simple, but the impact is strong. A drink consisting of a tablespoon of vinegar in a tall glass of water, drunk a few minutes *before* eating something sweet, flattens the ensuing glucose and insulin spikes. With that, cravings are curbed, hunger is tamed and more fat is burned. This is a very cheap trick, too: a standard bottle of vinegar costs under £2 at your corner store and contains over 60 one-tablespoon servings. You're welcome.

Vinegar is a sour-tasting liquid made by fermenting alcohol, thanks to common bacteria which turn it into acetic acid. These bacteria are ever present in our world – they are even in the air we breathe. If you leave a glass of wine sitting out on your table and go on holiday, when you return in a few weeks it will have turned to vinegar.

Common varieties of vinegar include rice vinegar, white wine vinegar, red wine vinegar, sherry vinegar, balsamic vinegar and apple cider vinegar (ACV), with the latter being most popular for this hack. The reason is that most people find that it tastes better than the other vinegars when diluted in that tall glass of water. But all vinegars work identically on our glucose, so pick the one you please. (Note that lemon juice does not have the same effect because it contains citric acid, not acetic acid.)

## Meet Mahnaz

Vinegar has been touted as a health remedy for centuries. In the 18th century it was even prescribed in tea form to people with diabetes. In Iran it's consumed many times a day, in various water-based drinks, by people of all ages. 'In my family we've been drinking apple cider vinegar for generations,' explained Mahnaz, a Glucose Goddess community member from Tehran. 'My grandmother makes her own and distributes it to all family members. We drink it because it's part of our culture and it's been passed on that it's healthy. As to why exactly it's good for us, I had no idea until I found your account.'

This is Mahnaz's grandmother's recipe, in case you want to get into fermenting, too:

Mash clean and sweet apples.
Put into barrels.
Cover and leave alone for 10 to 12 months.
The place should be hot.
And sunlight is very good.
Insects are okay and a sign of good vinegar.
So don't panic, they are just helping.
When it is ready, you should strain the liquid very well, twice, using a fabric with tiny holes.

Though people have been drinking vinegar for centuries,

it's only recently that scientists have been able to understand the mechanisms behind its health benefits.

In the past decade, a couple of dozen research teams around the globe have measured the effects of vinegar on our body. Here is how most of the studies went: put together a group of 30 to a few hundred participants. Ask half the group to drink one or two tablespoons of vinegar in a tall glass of water before their meals for three months and give the other group a placebo, something that tastes like vinegar but isn't vinegar. Track their weight, blood markers and body composition. Make sure both groups maintain identical diet and exercise, sit back, grab popcorn and watch.

What researchers found was that by adding vinegar before meals for three months, the subjects lost 2 to 4 pounds and reduced their visceral fat, waist and hip measurements and triglyceride levels. In one study, both groups were put on a strict weight loss diet, and the vinegar group lost twice as much weight (11 versus 5 pounds), even though they ate the same number of calories as the non-vinegar group. A Brazilian research team explained that because of its effect on fat loss, *vinegar is more effective than many thermogenic supplements touted as fat burners.*

The positive effects of vinegar are many. In people without diabetes, in people who have insulin resistance, in people with type 1 diabetes or type 2 diabetes alike, as little as one tablespoon a day significantly decreases glucose levels. The effects are also seen in women with PCOS: in a tiny study (which definitely needs replication before it's confirmed), four out of seven women got their periods back in 40 days when they added one vinegar drink a day.

Here's what happened in all these participants' bodies: when they drank vinegar before eating a meal rich in carbo-hydrates, the glucose spike from that meal was reduced by 8 to 30 per cent.

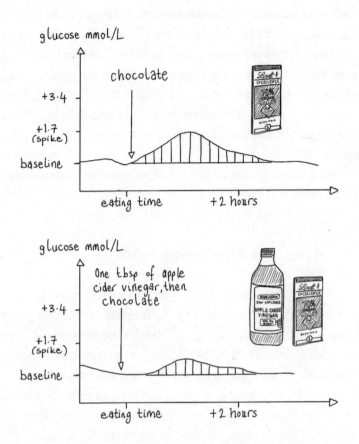

**Here is a test I ran to illustrate the science: vinegar curbs a glucose spike.**

To understand how this happens, we have an important clue: *the amount of insulin also decreases when vinegar is consumed before eating* (by about 20 per cent in one study).

This tells us that drinking vinegar does not flatten glucose curves by increasing the amount of insulin in the body. And this is a very good thing. Indeed, you *could* flatten a glucose curve by injecting someone with insulin or giving them a medication or a drink that would release more insulin into their system. This is because the more insulin there is in the body, the more your liver, muscle and fat cells work to remove any excess glucose from the bloodstream and quickly store it away. However, although insulin brings glucose levels down, it also increases inflammation and weight gain. What we really want to do is flatten our glucose curves *without increasing the amount of insulin in the body*. Which is what vinegar does.

So how does it work? Scientists believe that several things could be at play.

## How vinegar works

Remember the enzyme that Jerry and humans have in common, alpha-amylase? This is the enzyme that in plants chops starch back up into glucose and in humans turns bread to glucose in our mouths. Scientists have found that the acetic acid in vinegar temporarily inactivates alpha-amylase. As a result, sugar and starch are transformed into glucose more slowly, and the glucose hits our system more softly. You may recall from Hack 1, 'Eat foods in the right order', that fibre also has this effect on alpha-amylase, which is one of the reasons fibre helps flatten our glucose curves, too.

Second, once acetic acid gets into the bloodstream, it penetrates our muscles: there, it encourages our muscles to make glycogen faster than they usually would, which in turn leads to more efficient uptake of glucose.

These two factors – glucose being released into the body more slowly and our muscles uptaking it more quickly – mean that there is less free-flowing glucose present, so less of a glucose spike.

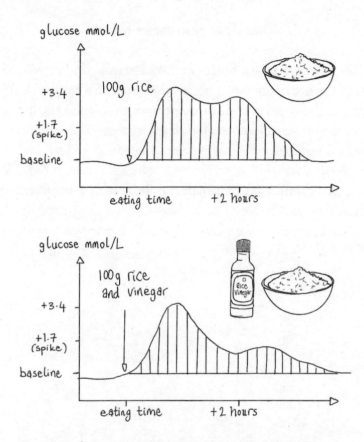

**Any type of vinegar works. One tablespoon of rice vinegar in a bowl of white rice (as per Japanese tradition) will help steady your glucose levels.**

What's more, acetic acid not only reduces the amount of insulin present – which helps us get back to fat-burning mode – it also has a remarkable effect on our DNA. It tells our DNA to reprogramme slightly so that our mitochondria burn more fat. Yep. For real.

# What does this mean for us?

This hack works for both sweet and starchy foods. Maybe you're ready to dig into a big bowl of pasta. Or maybe you're at a birthday party and you must have chocolate cake in the middle of the afternoon. Reach for vinegar first to counter some of the side effects of a glucose spike.

Grab a tall glass of water, and pour a tablespoon of ACV into it. If you don't like the taste, start with a teaspoon or even less, and build up to it. Grab a straw, down the drink either less than 20 minutes before, during the course of, or less than 20 minutes after eating the glucose-spiking food.

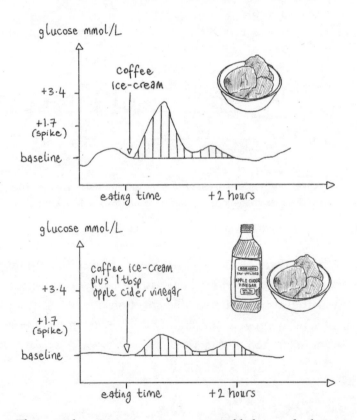

**There you have it: eat your ice cream, and help your body, too.**

Here's an even easier way to use this hack: now that you're adding a green starter to all your meals, you can add some vinegar to your dressing. In the first-ever study looking at vinegar and glucose spikes, two meals were consumed: one group ate a salad with olive oil, then bread, and the other group ate a salad with olive oil and vinegar, then bread. In participants who had the dressing with vinegar, the glucose spike was 31 per cent smaller. So order vinaigrette instead of a mayonnaise-based dressing next time.

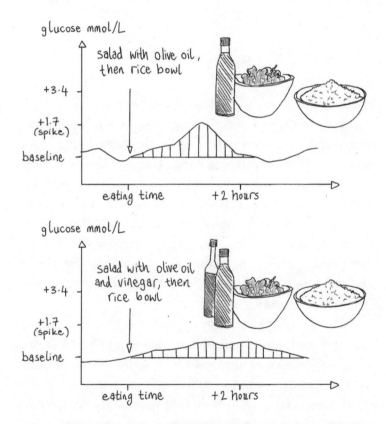

**For your green starter, the best dressing for your glucose levels incorporates vinegar – like a traditional vinaigrette.**

Vinegar to curb a glucose spike is most useful when it's consumed during a meal that would otherwise cause a big spike, but really, you can use it anytime – depending on your level of commitment. (And in the next few pages, I share even more recipes for using vinegar in this way.)

To be clear: you can't vinegar yourself out of a bad diet. Vinegar curbs spikes, but doesn't erase them. It will help if you add it to your diet – but it's not a justification to eat more sugar, because on balance, that would make your diet worse than before.

## Back to Mahnaz

Mahnaz's mother was diagnosed with type 2 diabetes after her third pregnancy 16 years ago. It was hard for her to manage that condition, despite the family apple cider vinegar production (consuming vinegar alone will not prevent someone from getting diabetes). So Mahnaz told her about the hacks in this book. Mahnaz's mother started to eat her food in the right order and switched to savoury breakfasts. She was already drinking vinegar in a tall glass of water, so she kept at it. In four months, her fasting glucose level went from 11.1 mmol/L to 6.1 mmol/L, from severely diabetic to diabetic no longer.

I mention this in part to remind you that the hacks in this book are tools in your toolbox. Some may be easier than others to incorporate into your life. Some may work better for you than others and in different combinations. But they are all beneficial. And the more of them you use, the more successful you'll be in flattening your glucose curves.

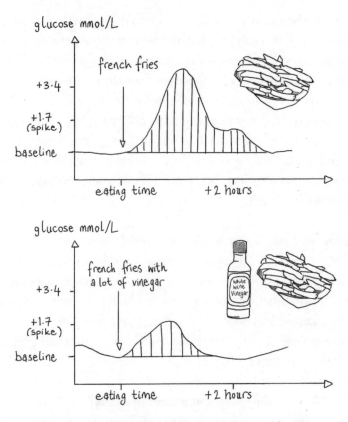

**All vinegars work. Here, white wine vinegar. The Brits had it right!**

### Why do I need a straw?
Even though diluted vinegar is not acidic enough to damage your teeth's enamel, I would suggest you drink it with a straw just to be safe. Never swig it straight from the bottle. As part of other foods, such as vinaigrette, it's fine.

### Are there any negative side effects?
You shouldn't experience any negative side effects as long as you stick to drinkable vinegar – i.e., vinegar with 5 per cent acidity (cleaning vinegar has 6 per cent acidity, so if it's

next to the mops and toilet paper at the supermarket, don't drink it!). For some people, vinegar can irritate their mucous membranes; for others, it can cause heartburn. It's not recommended for people with stomach disorders, although that is just a precaution – no studies have been done to measure the effects. Vinegar does not appear to damage the stomach lining, since it's actually less acidic than gastric juices – and even less acidic than Coca-Cola or lemon juice. Again, it's up to you – listen to your body and if vinegar doesn't agree with you, don't force it.

### Is there a limit to how much I can drink?

Well, yes. A 29-year-old woman who consumed 16 tablespoons of vinegar every day for six years was admitted to a hospital because of very low levels of potassium, sodium and bicarbonate. So don't do that. That's way too much. But most people do fine drinking a tablespoon in a tall glass of water a few times a day.

### Can I have it while pregnant or breastfeeding?

Most standard vinegars are pasteurised and safe to consume. Apple cider vinegar, on the other hand, is usually unpasteurised, which may present risks to pregnant women. Check with your doctor first.

### Uh-oh, I forgot to drink vinegar, and now I've eaten a slice of cake. Is it too late?

No! I do this all the time. Sometimes the slice of cake is so appealing that I forget about the pre-cake drink. No worries. Drinking it after eating something sweet or starchy (again, up to 20 minutes after) is much better than not drinking it at all. It still has glucose-lowering effects.

### How about pills and gummies?

When it comes to vinegar pills or capsules, the jury is still out. They may work as well as vinegar in liquid form, but it's not certain. If you want to try pills, you might need to swallow three or more to get the amount of acetic acid in a tablespoon of vinegar (about 800 milligrams).

Gummies aren't a good move: they contain sugar (about 1 gram of sugar per gummy). So not only might they not work to flatten your glucose curves, they could actually lead to spikes. (I reached out to one leading apple cider vinegar gummy brand to ask for scientific backing for its claims – I didn't get an answer.)

### How about kombucha?

Kombucha has less than 1 per cent acetic acid, and if it's not homemade, it often has added sugar in it. So although it is not a spike slasher, it still has some health benefits: since it's a fermented food, it contains beneficial bacteria that fuel the good microbes in our gut.

### I don't like the taste of vinegar. What should I do?

Start with a small quantity, and work your way up. Or try white vinegar instead of apple cider vinegar (some people prefer the taste). Or you could consider mixing the vinegar with some other ingredients – it doesn't matter what you add (except don't add sugar, as that will negate the effects).

Here are some recipes from Glucose Goddess community members:

- A cup of hot cinnamon tea and 1 tablespoon of apple cider vinegar
- A glass of water, a pinch of salt, a pinch of cinnamon and 1 teaspoon of apple cider vinegar
- A glass of water, a pinch of salt, 1 teaspoon of liquid

aminos and 1 tablespoon of apple cider vinegar
- A teapot of hot water, with a wedge of lemon, some ginger root, 1 tablespoon of apple cider vinegar, and a pinch of allulose, monk fruit, stevia extract or erythritol for sweetness
- Sparkling water, ice and 1 teaspoon of apple cider vinegar
- Vegetables fermented in a jar full of apple cider vinegar

## Let's recap

Adding vinegar to our diet, either in a drink or in a salad dressing, is an excellent way to flatten our glucose curves. It does so in two ways: it slows the arrival of glucose in the bloodstream, then increases the speed at which our muscles soak it up and turn it into glycogen. Speaking of muscles, they seem to be pretty good at this job...

# HACK 8: AFTER YOU EAT, MOVE

Every three to four seconds, our eyelid muscles receive a message from our brain in the form of electric signals, or *impulses*. The signals contain a simple instruction: 'Blink now, please, so that we can hydrate these eyes and keep reading this amazing book.' Across our entire body, muscles contract to make us walk, lean, grab, lift and more. Some muscles we consciously control (for example, our fingers), others we don't (for example, our heart).

The more and the harder a muscle is told to contract, whether consciously or unconsciously, the more energy it needs. The more energy it needs, the more glucose it needs. (The mitochondria in muscle cells can also use other things to make energy, such as fat, but when glucose is abundant, that's the fast, ready fuel of choice.) There is a special name, by the way, for the energy created out of the ashes of glucose to fuel our cells: adenosine triphosphate, or ATP.

The rate of glucose burning varies widely depending on how hard we're working – i.e., how much ATP is required by our muscles. It can increase 1,000-fold from when we are at rest (sitting on our couch watching TV) to when we are intensely exercising (sprinting to catch our dog running across the park).

With every new muscle contraction, glucose molecules are burned up. And we can use this fact to our advantage to flatten our glucose curves.

# Meet Khaled

Khaled is 45 years old. He lives in the sunny, hot United Arab Emirates, where year-round beach days are the norm. Up until recently, Khaled didn't lie out to tan when he went to the seaside – he always wore a T-shirt, he said, to hide his belly from his friends.

Change is hard, so the best shot we have at making it work is choosing strategies that require very little effort but lead to big results. (Like, say, the hacks in this book.)

Like many of us, totally understandably, Khaled had no desire to change what he ate, but he was open to other ideas. Right before the Covid-19 pandemic, he came across the Glucose Goddess Instagram account. Seeing the effect of these hacks, mapped out in graphs, ignited something inside him – not least because his father and siblings have diabetes. When lockdown set in, Khaled suddenly had a lot of time on his hands, and he decided to try something new – as long as it was easy.

He decided he'd try walking after meals, which is one of the hacks I talk about on my Instagram account. Nothing he ate had to change. After his lunch of rice and meat, he just had to go for a 10-minute walk in his neighbourhood. As he walked, he imagined the glucose from the rice moving into his leg muscles, instead of heading to his fat reserves. When he got home, he surprised himself – instead of wanting to reach for sweets and then take a nap, as he usually did after lunch, he returned to his desk and worked through the afternoon. He felt… good. The next day, the 10 minutes of walking turned into 20. He continued with this new habit.

There are multiple traditions that recommend walking after eating, such as the Indian custom of '100 steps after a meal', and they exist for good reason. As soon as the influx of glucose (from a large bowl of rice, for example) hits our body, two things can happen. If we stay sedentary as the spike reaches its peak, glucose floods our cells and overwhelms our mitochondria. Free radicals are produced, inflammation increases and excess glucose is stored away in the liver, muscles and fat.

If, on the other hand, we contract our muscles as the glucose moves from our intestine to our bloodstream, our mitochondria have a higher burning capacity. They aren't overwhelmed as quickly – they are thrilled to use the extra glucose to make ATP to fuel our working muscles. On a continuous glucose monitor graph, the difference is stark.

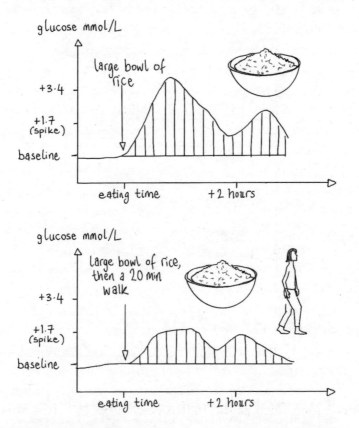

**When we eat starches or sugars, we have two choices: either stay still and let the spike happen, or move and curb the spike.**

Here's another way to think about it: when we exercise (again, just 10 minutes of walking is helpful), we make the fire on our grandfather's steam train bigger and hotter. Our grandfather shovels coal at a greater speed; the steam train burns it at a greater speed. Instead of accumulating, the extra glucose is used up.

So we can eat the exact same thing, and then, by using our muscles afterwards (within 70 minutes of eating; more on that below), flatten the glucose curve of that food.

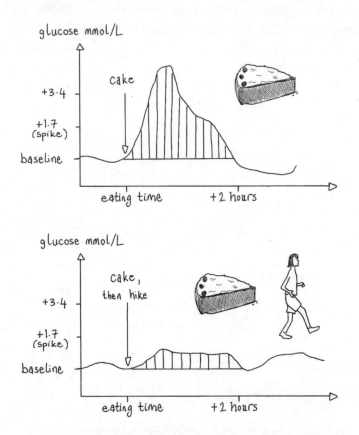

**If we sit on a chair for an hour after eating cake, the glucose will accumulate in our body and cause a spike. If we exercise instead, the glucose will almost immediately be used up by our muscles. It won't accumulate and cause a spike.**

Over the next six months, Khaled kept on walking for 20 minutes after lunch or dinner. Then he started to eat food in the right order. He lost 16 pounds. Remarkable, I know. And he's beaming.

He shared, 'I feel younger than ever before. When I compare myself to other people my age, I am doing way more, have more energy and am happier. My friends ask me what I've done… I'm happy to share the hacks. It's helped everyone in my family, too.'

Many people, like Khaled, walk 10 to 20 minutes after meals, and they've seen excellent results. A large 2018 research review looked at 135 people with type 2 diabetes and found that aerobic exercise (walking) after eating decreased their glucose spike by between 3 and 27 per cent.

If you want to hit the gym after meals, that's going to help even more – although some people find strenuous exercise on a full stomach quite hard. The good news is, you can work out at any time up to 70 minutes after the end of your meal to curb a glucose spike; 70 minutes is around the time when that spike reaches its peak, so using your muscles before that is ideal. You can also use your muscles acutely in a push-up, a squat, a plank, or any weight-lifting exercise. Resistance exercise (weight-lifting) has been shown to decrease the glucose spike by up to 30 per cent and the size of further spikes over the following 24 hours by 35 per cent. It's rare that you'll be able to curb the *entire* glucose spike, but you can make a sizeable dent in it.

And here's the kicker: when we move after eating we flatten our glucose curve *without increasing our insulin level* – just as was the case with vinegar. Although our muscles usually need insulin to stash glucose away, if they are currently contracting, *they don't need insulin to be able to uptake glucose.*

And the more our muscles contract and uptake glucose without needing insulin, the smaller the glucose spike will be, so the less insulin will be dispatched by the pancreas to deal with the remaining glucose. This is great news all around. Going for even just a 10-minute walk after a meal will lessen the side effects of whatever we just ate. And the longer we work out, the more our glucose and insulin curves will flatten.

## What to do while you watch TV after dinner

You're home, you've had a bowl of pasta for dinner (after a green salad, right?), you're about to sit on the couch and turn on your favourite TV show. But if you're able to multitask, try doing some squats while you're watching the screen. Or try a wall sit with your back against the wall, do triceps dips off the edge of the couch, hold a side plank, or a boat pose on the carpet.

A Glucose Goddess community member named Monica has a fun set-up: she keeps a kettlebell behind her couch, and after eating something sweet, she sets a 20-minute timer on her phone – when it goes off, she grabs the weight and does 30 squats, holding the kettlebell.

Office variation: you have no time to go for a walk after your meal. That's okay. Go up and down the building stairs a couple of times, pretending that you need to use the bathroom. While in a meeting, do some quiet calf raises on the floor. Or a set of elevated push-ups against your desk. Problem solved.

TRY THIS: Rate how you feel when you have a sweet
snack and then stay sitting. Rate how you feel when you
have the same treat and walk 20 minutes afterwards.
How is your energy? How is your hunger level
in the next few hours?

### How quickly after eating should I exercise?

Monica gets active 20 minutes after eating, but you can
exercise anytime within 70 minutes after eating to see an
effect. As mentioned above, you want your muscles to start
contracting before the glucose spike reaches its peak. I like
to go for a walk or do strength or resistance exercises in front
of the TV about 20 minutes after a meal. But in various
studies, many different scenarios have been tested: some
people started walking right after they put their fork down,
some 10 to 20 minutes after finishing their meal. Others
waited 45 minutes after eating to start a workout. But all
worked well.

### Should I exercise before or after a meal?

Exercising after a meal seems to be the best option, but
before is also useful. In a study of resistance training in obese
people, exercising *before* dinner (eating 30 minutes after the
workout was over) lowered their glucose and insulin spikes
by 18 per cent and 35 per cent, respectively, as opposed to
30 per cent and 48 per cent if the exercise was started 45
minutes *after* dinner.

### How about other times during the day?

Exercising *anytime* is great for you. And it has many more
positive side effects than just curbing a glucose spike. Among
other things, it improves our mental well-being, energises us,
helps our heart stay healthy and reduces inflammation and

oxidative stress. Whether you're fasted or not, if you take up a new physical activity, your overall glucose levels will start decreasing as you put on muscle mass.

However, if you're thinking of adding more walking to your daily regime and you can do it at any time, it'll be more impactful after meals.

### How many minutes of exercise do I need?

It's up to you to find what works. Studies usually look at 10 to 20 minutes of walking or 10-minute strength or resistance sessions. I've found that I have to do about 30 squats to see any change to my glucose level.

### Why does fasted exercise lead to a glucose spike? Is that bad?

When you exercise and you haven't yet eaten, i.e., you are engaging in fasted exercise, your liver releases glucose into your blood to fuel the mitochondria in your muscles. This shows up on a glucose monitor as a spike – because there is one. These spikes do cause oxidative stress, by increasing free radicals, but the exercise that causes them also increases your ability to get rid of free radicals, and, importantly, that improved defence against free radicals lingers longer than the acute, exercise-induced production of free radicals. Thus, the net effect of exercise is to reduce oxidative stress. Exercise is therefore considered a *hormetic* stress on the body. This means that it is a type of beneficial stress that causes our bodies to become more resilient.

## Let's recap

If you're going to eat something sweet or starchy, use your muscles afterwards. Your muscles will happily uptake excess glucose as it arrives in your blood, and you'll lessen the glucose spike, reduce the likelihood of weight gain, and avoid an energy slump. Post-eating sleepiness is particularly curbed by this technique. And it works even better when you drink vinegar mixed into a tall glass of water before you eat.

Now you know the amazing combo for snacking on something sweet without incurring a big glucose spike in your body: vinegar before, exercise after.

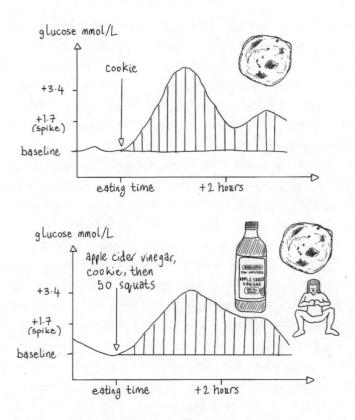

**The more the merrier: combining hacks is incredibly powerful. Bookending a sweet treat by drinking vinegar beforehand, then using your muscles afterwards, will help you reduce the side effects.**

# HACK 9: IF YOU HAVE TO SNACK, GO SAVOURY

I've mentioned how glucose impacts both our bodies and our minds throughout this book. Back when I embarked on this research, however, it was always easier to discern glucose's physical rather than its mental effects. I knew why I saw acne on my nose or why I gained weight. Until one day – when I looked at the data from my own glucose monitor after I ate a doughnut.

Since my accident at 19, I've struggled with a mental health condition that I've named 'splitting' or 'feeling split'. Clinically, it's called *depersonalisation*. When it happens, it feels as though I partially leave my body. When I look in the mirror, I don't recognise myself. When I look down at my hands, I think they belong to someone else. A fog sets in before my eyes. I lose my unified sense of 'I', and my mind starts to spin uncontrollably as I think about existential questions. It's very scary, especially if I'm alone.

The thing that gets me through those moments is remembering that they will pass. I've found a lot of help in talk therapy, eye movement desensitisation and reprocessing, or EMDR (where I remember the incident while my therapist taps on my knees alternately), and craniosacral therapy (a form of bodywork).

I was lucky that I knew someone close to me who had experienced the same thing when he was younger – my cousin. I would text him whenever I needed reassurance. 'I know it's awful. Trust me, it will pass,' he would respond. I also turned to my diaries. I wrote a lot.

I felt split for a full year after my surgery. Then the feeling would come and go once a week or once a month and last for a few hours. I did everything to try to find out what triggered and what resolved that feeling. But most of the time, I simply didn't know.

Then, eight years after my accident, I realised that one of the triggers could be... food.

In April 2018, my boyfriend and I and a couple of friends were visiting the seaside city of Kamakura on the coast of Japan. I had been wearing a glucose monitor for just about a month.

We had breakfast very early. Five hours later, we were hungry again. We stopped for coffee and doughnuts, then headed for a walk by the ocean.

As we talked about our next adventures – viewing cherry blossoms, visiting Harajuku – I began to notice a shift in my mental state. The feeling was all too familiar. I knew I was about to split.

The fog set in. I looked at hands that were not my own. I knew I was speaking, but I didn't really know what I was saying or why. As is usually the case, I didn't confide in my friends, fearful that I would be a burden to them.

Through the fog, I scanned my glucose monitor. It was a habit by now; I had been doing it every few hours since I had started wearing it.

The doughnuts we had eaten 30 minutes before had created the biggest glucose spike I had ever seen: from 5.4 mmol/L to 10 mmol/L.

I realised I had potentially found a trigger for the splitting: a very steep glucose spike. And in fact, in the ensuing months and years, I was able to prove it. When I felt split, I would recall what I had eaten that day. It happened if I had

had chocolate cake for dinner instead of a normal meal or cookies for breakfast.

Now, this is not to say that flattening my glucose curves has cured my depersonalisation. I still feel split when I don't spend enough time alone, when I'm holding stress in my body, and for other reasons that I don't yet understand, and sometimes I'll experience a big glucose spike and not feel split at all. But this new awareness has definitely helped.

I did some research and didn't find any studies showing an episode of depersonalisation being triggered by food. I did discover, however, that in people with this mental health condition, some areas of the brain are more metabolically active – that is, they consume more glucose – than others. More glucose in the body, more glucose in the brain, so potentially more glucose in those hyperactive areas, too. Maybe that would cause the issue.

Certainly, we know that food affects how we feel. Science tells us that when people eat a diet that leads to lots of glucose spikes, they report worsening moods over time and more depressive symptoms than if they eat a diet of similar calories but with flatter curves.

Many community members have also shared that sugary foods increase their anxiety.

We all, from time to time, feel the urge to snack on something sweet – often when we're feeling sleepy. However, the idea that eating something sweet will energise us is a myth. A sweet snack doesn't give us more energy than a savoury snack, and it can actually just make us more tired shortly afterwards. Which, if you have to drive twelve hours a day, like Gustavo, can be positively dangerous.

# Meet Gustavo (again)

Gustavo taught us his wonderful broccoli-before-the-steakhouse tip that allowed him to enjoy dinners with his friends and flatten his curves. He's back, live from Mexico, with another piece of information.

Gustavo needs to drive long shifts between states for his sales job. He often stays up to 12 hours on the road at a time. In the old days, when he stopped at a petrol station, feeling exhausted, he would pick up some sweets or a granola bar to 'get some energy'. He'd get back behind the wheel, feel energised for about 45 minutes, then exhausted once again. He was very likely not metabolically flexible: his body couldn't switch to using his fat reserves for fuel, so he needed to eat starch or sugar often. Little did he know that, as we learned in Hack 4 ('Flatten your breakfast curve'), because of how insulin works, the glucose in sweets or a granola bar tends to go to storage rather than to be used as fuel. So when we eat something sweet, there is actually *less* energy circulating in our body after digestion than when we eat something savoury. Gustavo would briefly feel perked up by his snack. But it didn't last very long, and an hour later he'd be tired and have to stop for another one.

As I mentioned in Hack 2, Gustavo first decided to make lifestyle changes after people close to him passed away from complications related to type 2 diabetes. Gustavo also replaced his cereal in the morning with a glucose-steady smoothie that he makes with flaxseed, nopal (prickly pear cactus) and maca root (he says it tastes better than it sounds). Out with the sitting around after eating, in with the walking. Now it was time to smarten up his snacking when on the road: no more sweets or granola bar from the petrol station; much better to take some carrots, cucumber and peanut butter with him. Which he now always does.

These days, with his flatter glucose curves, Gustavo doesn't feel the overwhelming desire to nap in the middle of the highway. His energy is steady throughout his drives. He has also lost 88 pounds, reduced his depression medication and cleared his brain fog.

If you're looking for energy, and I know it's counter-intuitive, opt for a savoury snack rather than a sweet one. And not a starchy one, either, since starch also turns to glucose.

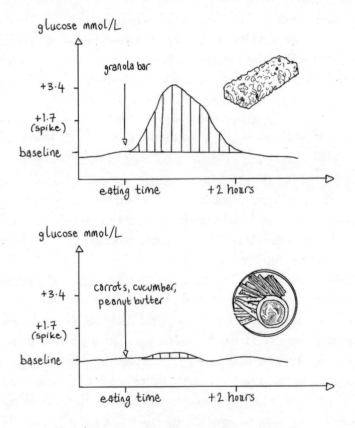

**For steady energy, pick snacks that don't spike your glucose levels.**

# The 30-second no-glucose-spike savoury snack

Here are my go-to savoury snacks:

- A spoonful of nut butter
- A cup of 5 per cent Greek yoghurt topped with a handful of pecans
- A cup of 5 per cent Greek yoghurt with nut butter swirled into it
- A handful of baby carrots and a spoonful of hummus
- A handful of macadamia nuts and a square of 90 per cent dark chocolate
- A hunk of cheese
- Apple slices with a hunk of cheese
- Apple slices smeared with nut butter
- Bell pepper slices dipped into a spoonful of guacamole
- Celery smeared with nut butter
- A handful of pork rinds
- A hard-boiled egg with a dash of hot sauce
- Lightly salted coconut slivers
- Seeded crackers with a slice of cheese
- A slice of ham
- A soft-boiled egg with a dash of salt and pepper

# HACK 10: PUT SOME CLOTHES ON YOUR CARBS

I don't know about you, but I don't always have the time to sit down for a meal. And often I'm hungry when I'm on the go and there is no source of healthy food in sight – just a corner shop near my next meeting or a cafe at my airport gate as I head for a flight. So this hack is for those times – for real-life eating when we have to grab something on the way to the bus, when we're at a party or a business breakfast. It's for those times when we're going to eat a slice of cake for breakfast because we're hungry and it's there.

The solution is simple, and I've mentioned it throughout these pages: combine starches and sugars with fat, protein or fibre. Instead of letting carbs run around naked, put some 'clothes' on them. Clothes on our carbs reduce how much and how quickly glucose is absorbed by our bodies. Have the brownie at your friends' place, but ask for Greek yoghurt with it, too. Have the bagel at the business meeting, but choose the one with smoked salmon in it. Buy a take-out lunch, but add to it ingredients from the nearest food shop: cherry tomatoes and some nuts. The same goes for when you are at home: if you're making cookies, add nuts to them; if you're serving apple crumble, offer cream to go on top.

When you do enjoy carbs (and you will and should and must), make it a habit to add fibre, protein or fat and, if you can, to eat those first. Even savoury snacks – which are already better for your glucose curves but may still contain starch – should have clothes on: add avocado and cheese to toast, spread nut butter on rice cakes and eat some almonds before your croissant.

### I've heard that adding fat to a meal is bad because it increases the insulin spike.

This belief was popularised by a Frenchman, Michel Montignac, in the 1980s. But the most recent science demonstrates otherwise. Adding fat to a meal does not increase the insulin spike that meal causes. It does not tell our bodies to secrete more insulin. In fact, eating fat before a carb-rich meal *decreases* the amount of insulin produced in response to the meal.

Eating carbohydrates alone isn't just bad for our glucose levels, it also plays havoc with our hunger hormones. So we go from feeling full to being hungry again very quickly.

By putting clothes on our carbs, we avoid hunger pangs. We also avoid being *hangry*, something I experienced almost every day when I was a teenager. This is because when we eat carbohydrates on their own, ghrelin, a hormone that tells us to eat, fluctuates rapidly, then makes us hungrier than we were before eating.

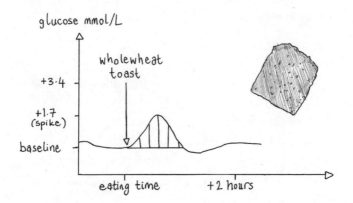

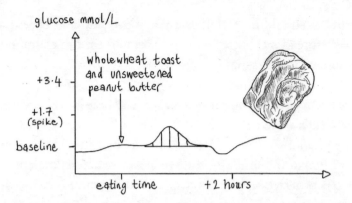

**Often, putting clothes on your carbs makes them taste better, too.**

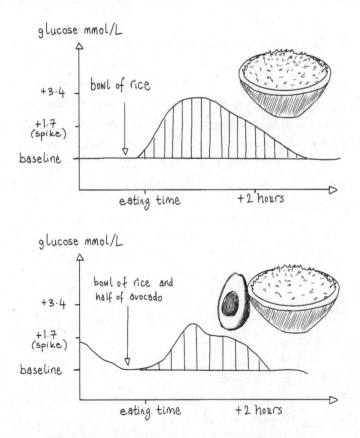

**Rice is better for our glucose when it has clothes on.**

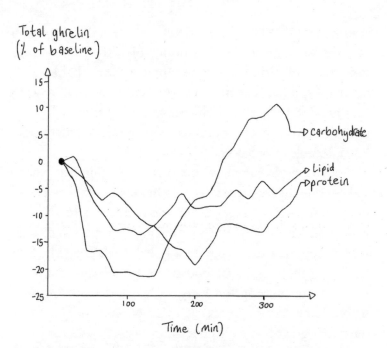

**When we eat carbohydrates on their own, ghrelin, a hormone that tells us to eat, fluctuates rapidly, then makes us hungrier than we were before eating. Carbs drive our hunger up and down like a rollercoaster, while fats and proteins don't.**

## Meet Lucy and her temper

'I was concerned I would destroy all my relationships, one by one.' This confession came out of the mouth of Lucy, aged 24, a heptathlon athlete who lives in the United Kingdom. Lucy was snapping at her parents and saying mean things to her friends. Her behaviour was turning her into someone no one wanted to be around. As she came to discover, she wasn't to blame – naked carbs were.

Thousands of scientific studies show how glucose spikes harm our body, but as I mentioned in the previous hack, evidence of the connection between glucose and the mind is still emerging. I've already described the research proving that the more glucose spikes in our diet, the more we feel symptoms of depression and anxiety. But thanks to a fascinating recent experiment, we also know that when we eat a breakfast that causes glucose spikes, we're more likely to want to punish those around us – we become vindictive and less agreeable to our peers.

Lucy's confession may seem extreme, but her glucose spikes were also extreme. This is because Lucy has type 1 diabetes. People with type 1 diabetes don't have the ability to make enough insulin. Without insulin, when a spike comes in, the glucose can't make its way into cells properly. So it stays very high in their bloodstream for a long time, while the cells are starving for energy. This causes huge issues – before Lucy was diagnosed at 15 years old, she didn't even have enough energy to lift a fork.

On the first day of her new life with a diagnosis of type 1 diabetes, nurses at the hospital gave Lucy a plate of (naked) pasta to eat. They then taught her how to inject insulin with a syringe into her abdomen. The insulin diffused through her entire body, helping the glucose from the pasta make its way to the cells and bring down the spike caused by it.

The nurses explained: eat carbs at every meal and inject insulin at every meal. The bigger the glucose spike from what you just ate, the more insulin needs to be injected. This may sound simple to a person without diabetes, but getting the dosage right is a science. You have to constantly calculate where your glucose levels will be in the next hour or so, always planning ahead to avoid dreaded highs and lows. Eating, napping, exercising, all turn into a maths problem. Huge spikes and huge drops are the name of the game for most people with diabetes. To give you an example, once diagnosed and using insulin, Lucy's glucose level would go up to 17 mmol/L, then down to 4 mmol/L, then back up to

14 mmol/L, and back down to 4 mmol/L on a daily basis. Remember, my biggest spike as a person without diabetes was from 5.4 mmol/L to 10 mmol/L from a doughnut on an empty stomach – and I felt the side effects keenly.

Lucy felt the side effects even more keenly. She woke up every morning feeling hungover. Whenever her glucose levels were high, she'd snap at her mother. She couldn't help it and regularly cried out of regret afterwards. Home was one thing – until her teammates started avoiding her at school, too.

In me, a relatively small spike (compared to what a person with diabetes can experience) can trigger brain fog and depersonalisation. In Lucy, spikes caused uncontrollable anger. She felt stuck, too. She thought, *I guess I have to live with this.*

Lucy started perusing the type 1 diabetic forums for advice on how to deal with her symptoms. Other people with type 1 diabetes were talking about flattening their glucose curves and linking to my Instagram account.

Lucy found a few things there that helped her: first, she saw that a person without diabetes, like myself, can also experience glucose spikes into the 10s. That was shocking to her. She had always thought that for people without diabetes, glucose levels remained stable between 4 and 5 mmol/L all day long. That made her feel less alone: it's hard for *all of us* to flatten our glucose curves.

Second, she saw I was wearing a glucose monitor. She said, 'Seeing you wearing it proudly even though you don't need it, gave me the courage to put one on too. It helped me not be embarrassed.'

Finally, she saw that depending on what you eat, you really can flatten your glucose curves. Lucy understood that she could do something about how poorly she felt, body, mind and soul.

She met with her endocrinologist and made a plan (when you inject insulin or are on any type of medication, it's very important to talk to your doctor before changing the way you eat – to make sure you're not causing interactions that could be dangerous).

Lucy had always been told to eat carbohydrates at every meal – and especially at breakfast. The first thing she did, with her endocrinologist's supervision, was to flatten her breakfast: from orange juice and croissants (which she didn't even like) to salmon, avocado and almond milk. She used to see a spike of up to 16 mmol/L after breakfast. Now her glucose levels stay virtually flat.

Breakfast was easy to change, as were lunch and dinner, but snacks were less so. Lucy gets very hungry in the middle of the day because she trains so much, and she loves to reach for a banana or a chocolate bar.

What did she learn to do? To put some clothes on her carbs: she added nut butter to the banana, and she ate a hard-boiled egg before the chocolate. (A tip from Lucy: hard-boil eggs in batches every week and keep them in your fridge.)

With these hacks, Lucy's HbA1c level (the measure of glucose variability) dropped from 7.4 to 5.1 in three months – 5.1 is a level common among many people without diabetes. She injects about one-tenth the insulin she did before. And she's about ten times happier than before.

When we put clothes on our carbs, the game of Tetris our body plays with glucose goes from level 10 to level 1. There is less oxidative stress, fewer free radicals, less inflammation. And less insulin. With flatter glucose curves, we feel better and our mood is more stable. These days, instead of feeling hungover, Lucy wakes up refreshed. It seems simple, but it's often the smallest things that are the most meaningful: she walks down to the kitchen with a smile on her face and, without snapping, asks her mother if she can make her a coffee. She doesn't get so angry any more. She doesn't cry afterwards if she snaps at her parents or her teammates – because she doesn't snap so often these days. Her relationships are back to where she wants them. Her steady glucose levels allow her, she said, to 'be the person she wants to be, just a nice person, and that's the most important'.

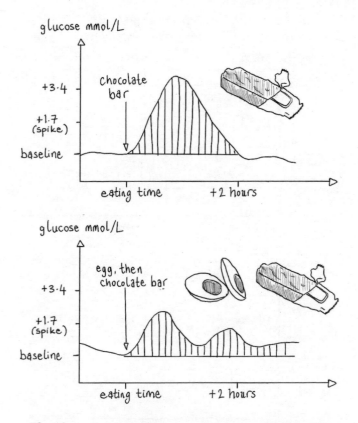

**If you're eating something sweet, put some clothes on it; fibre, fat or protein.**

I've heard a lot of stories like these. Flatter curves can lead to our being more patient with our kids, more loving with our partner and more supportive to our colleagues.

### What about fruit?

As I explained in Part I, the fruit we eat these days has been bred for centuries to contain more glucose and fructose and less fibre. So although whole fruit is still the healthiest way to eat sugar, we can go a step further and help ourselves some more by combining it with glucose-level-flattening friends – fat, protein and fibre.

Here are some tips to keep in mind:

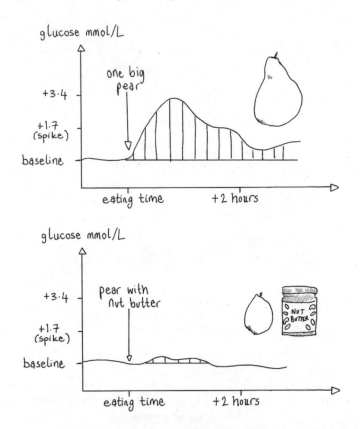

- Combine your fruit – top favourite partners in the Glucose Goddess community are nut butter, nuts, full-fat yoghurt, eggs and Cheddar cheese.
- Note that dried dates are some of the biggest

glucose bombs in the fruit kingdom. Yet they are said to help with managing diabetes. Go figure.

• And one last thing – when you have a choice between various fruits, the best option is berries. Tropical fruits and grapes are bred to contain high amounts of sugar, so eat them for dessert or put some clothes on them.

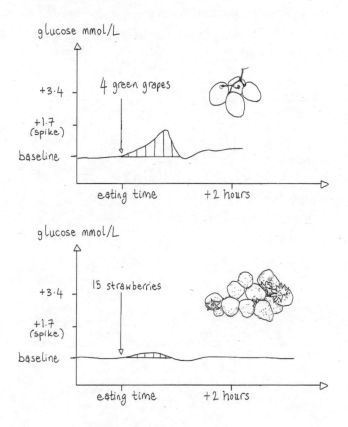

### Do whole grains still need clothes?

We often think that if grains are whole (brown rice, brown pasta, etc.), they are much better for us. The truth is, they are only very slightly better – starch is still starch. Pasta or bread that boasts 'whole grain' on its packaging has still been milled – which means that some of its fibre has gone. If you want bread that contains beneficial fibre, choose a very dark bread, such as seed bread or pumpernickel (as mentioned in Hack 2).

In the end, rice is still rice, even if it's whole-grain or wild rice. Don't let it go out naked. Mix in chopped fresh herbs, such as mint, parsley and dill, and roasted nuts, such as almonds or pistachios, and enjoy it alongside roasted salmon or chicken. *Voilà*, your carbs are dressed to the nines – and tastier to my mind, too.

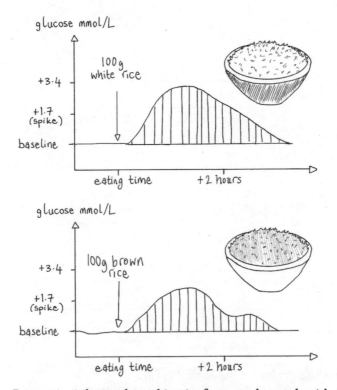

**Brown rice is better than white rice for your glucose, but it's still rice. Try to give it some clothes to flatten its curve.**

Lentils and pulses are different, though: they are better for you than rice, because although rice (or pasta or bread) is 100 per cent starch, lentils and pulses contain starch, fibre *and* protein. Remember: when we combine glucose with other molecules, whether we have diabetes or not, our body receives it at a more natural, manageable speed, and we curb the glucose spike.

---

**If you're eating carbs on their own ...**
Bread, corn, couscous, pasta, polenta, rice, tortillas, cake, sweets, cereal, cookies, crackers, fruit, granola, hot chocolate, ice cream, or anything else sweet

**... combine them with fibre, fat and/or protein:**
Any vegetable, beans, butter, cheese, cream, eggs, fish, Greek yoghurt, meat, nuts, seeds

---

### Which fat should I add?
Unlike sugar (there is no good or bad sugar; all sugar is the same, regardless of the plant it comes from), some fats are better for you than others.

Good fats are saturated (fat from animals, such as butter, ghee and coconut oil) or monounsaturated (from fruit and nuts such as avocados, macadamia nuts and olives). For cooking, use saturated fats – they're less likely to oxidise with heat. Monounsaturated fats, such as olive and avocado, can't stand the heat as well. A good rule of thumb to distinguish between them: cook with fats that are solid at room temperature when you can.

Bad fats (which inflame us, harm our heart health, make us gain visceral fat and increase our insulin resistance) are polyunsaturated and trans fats, which are found in processed oils – made from soybean, corn, rapeseed, safflower and rice bran oil – and fried foods, and fast foods. (The one seed oil that isn't as bad is flaxseed oil.)

We feel more satiated when there is fat in our diet, but we must be conscious of this dance: if we add tons of fat, our glucose spikes will be severely curbed, but we may start putting on weight. Add some fat, such as a tablespoon or two at a meal, but don't pour the entire bottle of olive oil onto your pasta.

Last, whenever you buy something, don't be fooled into thinking the 'low-fat' version is better for you: 5 per cent Greek yoghurt will help your glucose curves much more than a low-fat yoghurt will. (More on this in 'How to spot a spike on the packaging' on **page 232**.)

### How do I add fibre?
All veggies under the sun provide fibre. Along with nuts and seeds, they are the best clothes! You can even try fibre pills such as the ones made from psyllium husk.

### How do I add protein?
Protein is found in animal products, such as eggs, meat, fish, dairy and cheese, and also in many plant sources, such as nuts, seeds and beans. You can also use protein powders. Look for those with only one ingredient listed: the source of the protein. I usually choose hemp or pea protein powder. Make sure there are no sweeteners in it.

### I have type 1 diabetes. What should I do?
If you're going to change the way you eat in order to flatten your glucose curves, talk to your endocrinologist first. Adjusting your diet without adjusting your medication can cause unexpected highs and lows.

### I have type 2 diabetes. What should I do?

If you're currently insulin dependent or taking any medication, speak to your doctor before making any changes to your diet. With proper support, many people can reverse their type 2 diabetes. Many Glucose Goddess community members have shared their stories of how with me. For instance, Laura, who is 57 years old, started her journey towards flattening her glucose curves when she weighed 21st 6lb. She took metformin and glimepiride, two medications used to treat type 2 diabetes. After changing how she ate, thanks to what she learned on my Instagram account and working closely with her doctor, she lost 3st 6lb – and still counting – dropped her HbA1c level from 9 to 5.5, and reduced the dosage of her medications.

When I'm in Paris, where I live part of the time, I often go for a walk in the morning. At that time of the day, when I pass a bakery, I really want to tear into a baguette. When we are hungry, naked carbs look very appealing. But I keep in mind that the hungrier I am, the emptier my stomach is, the bigger the spike those naked carbs will cause. (This is why flattening our breakfast curve is so important.) I developed the habit of dressing up that baguette: these days I'll snack on almonds from the corner shop before my first bite of baguette, and when I get home, I'll spread some salted butter on it.

The hacks in this book have made a big difference in the lives of people in the Glucose Goddess community. I'm so excited for you to start trying them out, too. And as you do, remember: it's okay if you can't do them all the time. Even just adding them to your life a little bit, and when it's easy, will help your health.

# CHEAT SHEETS: HOW TO BE A GLUCOSE GOD OR GODDESS WHEN THINGS GET HARD

Here are a few tips based on specific situations that people have asked me for advice about: when a craving hits, when we're at a bar and when we're grocery shopping.

## When a craving hits

1. **Give a craving a 20-minute cooling-off period.** Back in hunter-gatherer days, decreases in our glucose levels signalled that we hadn't eaten in a long time. In response, our brain told us to choose high-calorie foods. Today, when we encounter a decrease in glucose levels, it's usually because the last thing we ate caused a glucose spike. Yet our brain tells us to do the same thing, to choose high-calorie foods, even though we aren't famished – we've got energy reserves. After a glucose drop, our liver quickly (within 20 minutes) steps in, releases stored glucose from those reserves into our bloodstream and brings our levels back to normal. At that point, the craving often dissipates. So next time you're about to grab a cookie, set a timer for 20 minutes. If your craving was due to a glucose drop, it will be gone by the time the alarm rings.
2. If the 20 minutes have come and gone and you're still thinking about that cookie, **set it aside for dessert at**

**your next meal.** In the meantime, consciously notice that you're having a craving and remind yourself that you've experienced this before and it will pass. Then try these craving killers: liquorice root tea or a spoonful of coconut oil swirled into coffee. Other things to try: peppermint tea, pickle juice, gum, a big glass of water with a big pinch of salt. Brush your teeth. Or take a walk.

3. If you can't wait for dessert at your next meal and you've decided you're going to go ahead and eat what you're

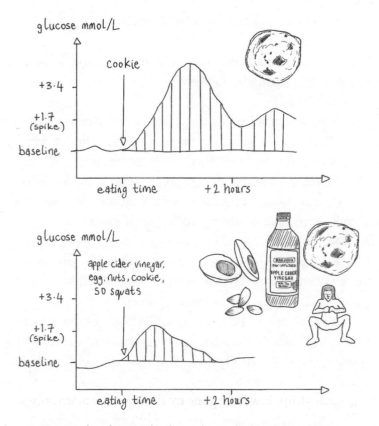

**Here is the ultimate hack combo to deal with a craving.**

craving right now, **drink a tall glass of water with a tablespoon of apple cider vinegar swirled in** (or as close to a tablespoon as you like).

4. **Then put some clothes on your carbs.** Have an egg, a handful of nuts, a couple of spoonfuls of 5 per cent Greek yoghurt, or a head of roasted broccoli before the thing you're going to eat.

5. **Eat that thing.** Enjoy it!

6. **Use your muscles and move within the next hour.** Go for a walk or do some squats. Whatever works for you.

## When you're at a bar

When you order a drink at a bar, you don't have to order a glucose and fructose spike along with it. (That is a lot for the liver to handle.)

Alcohols that keep our levels steady are wine (red, white, rosé, sparkling) as well as spirits (gin, vodka, tequila, whisky and even rum). We can drink these on an empty stomach, and they don't cause a glucose spike. Watch out for mixers: adding fruit juice, something sweet, or tonic will cause one. Drink your alcohol on the rocks, with sparkling water or soda water, or with some lime or lemon juice. When it comes to beer, which causes spikes because of its high carb content, ale and lager are preferable to stout (such as Guinness) and porter. Even better, go for low-carb beer.

And if you're nibbling on appetisers, go for the nuts and olives, as they will help balance your glucose levels. Try to stay clear of the crisps if you can, as they will cause a glucose spike.

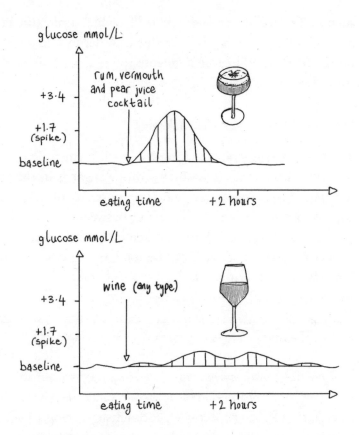

**Wine is fine, champagne and spirits, too; but stay away from cocktails and beer.**

## When you're grocery shopping

You'll naturally flatten your glucose curves if you cut back on processed foods, but for the occasions when you do buy them, here's what to keep in mind.

The items on supermarket shelves don't get a gold star for honesty. Far from it. If a processed food will cause a glucose spike, it's not going to be up front about it on the packaging. It's going to hide that secret, distracting you with labels such as 'fat-free' or 'no sugar added' – which unfortunately don't mean that the food is healthy. In order to find out if a processed food will cause a glucose spike, don't look at the front. Look at the back.

### How to spot a spike on the packaging

The first place to look is the ingredients list. Ingredients are sorted in descending order by weight. If sugar is in the top five ingredients, that means a hefty proportion of that food consists of sugar, even if it doesn't taste sweet – a soft white roll, for example, or a bottle of ketchup – and will cause a glucose spike. If sugar is in the top five ingredients, you know what that means: a hidden fructose spike.

---

**The many names of sugar on an ingredients list**

Look for these: agave nectar, agave syrup, barley malt, beet sugar, brown rice syrup, brown sugar, cane juice crystals, cane sugar, caramel, coconut sugar, icing sugar, corn syrup, corn syrup solids, crushed fruit, date sugar, dextrin, dextrose, evaporated cane juice, fructose, fruit juice, fruit juice concentrate, fruit purée concentrate, galactose, glucose, glucose syrup solids, golden sugar, golden syrup, grape sugar, high-fructose corn syrup (HFCS), honey, icing sugar, malt syrup, maltodextrin, maltose, maple syrup, muscovado sugar, panela sugar, pressed fruit, raw sugar, rice syrup, sucanat, sucrose, sugar.

---

A special mention goes to 'fruit juice', 'fruit juice concentrate', 'fruit purée concentrate' and 'pressed fruit': more and more, these pop up on yoghurt containers, and cereal and granola boxes. As you know, as soon as a fruit is denatured and processed and its fibre is extracted, it becomes sugar like any other sugar. When you pick up a juice or a smoothie, assess it as you would any other processed food: if the main ingredient is sugar – i.e. it is one of the 'fruit' by-products listed above – skip it. Eat a peach or an apple instead.

ingredients

half a pressed apple

half a crushed peach

13 pressed grapes

11 crushed raspberries

a dash of lemon juice

**The ingredients list on an Innocent smoothie: sugar under four different names (and a dash of lemon juice). I know they look cute, but remember, fruit juice is just sugar.**

**German sweets made with 25 per cent fruit juice (but the sugar from fruit juice is just the same as sugar from sugar beets).**

> **INGREDIENTS:** WHEAT FLOUR, SUGAR, VEGETABLE
> GLYCERIN, FRUCTOSE, DEXTROSE, MALTODEXTRIN,
> VEGETABLE AND MODIFIED PALM OIL SHORTENING,
> PALM KERNEL AND/OR PALM OIL, MODIFIED CORN
> STARCH, APPLE POWDER, PALM OIL, MODIFIED MILK
> INGREDIENTS, STRAWBERRY PUREE CONCENTRATE,
> CORN STARCH, BAKING POWDER, SOY LECITHIN, SALT,
> ACETYLATED TARTARIC ACID ESTERS OF MONO- AND
> DIGLYCERIDES, COLOUR (CARROT JUICE CONCENTRATE),
> SODIUM CITRATE, NATURAL FLAVOUR, CELLULOSE GEL,
> CITRIC ACID, MALIC ACID, MONO- AND DIGLYCERIDES,
> CELLULOSE GUM, SODIUM ALGINATE.
> **CONTAINS WHEAT, MILK AND SOY INGREDIENTS.**

The ingredients in Special K fruit crisp bars. Can you find the six different sugar names listed here?

## Stick to the facts

Sometimes it seems as though every single part of packaging is trying to confuse us. But I'm happy to announce that there is one haven of objective information: the Nutrition Facts.

One thing to keep in mind before we begin: in recent years manufacturers have been reducing the recommended serving sizes on their packaging to make things look better in terms of grams of sugar. A smaller serving size means less sugar per serving. But come on, who eats only *two* Oreos? So know that the absolute numbers you see on packaging are not the most important thing. Rather, it's the *ratios* that hold the key. Let me explain this powerful way to decode things.

First things first: you can skip right past the Calories line. Yes, it's the line in the biggest type, because that's what manufacturers want you to focus on. But as I've explained, the molecules matter more than the calories. And in the Nutrition Facts, the molecules in a food are spelled out for all to see – if you know where to look.

| Nutrition Facts | | |
|---|---|---|
| **Serving size** | | |
| **Amount Per Serving** | | |
| **Calories** | | **0** |
| | | **% Daily Value*** |
| **Total Fat** 0g | | **0%** |
|   Saturated Fat 0g | | **0%** |
|   *Trans* Fat 0g | | |
| **Sodium** 0mg | | **0%** |
| **Total Carbohydrate** 0g | | **0%** |
|   Dietary Fiber 0g | | **0%** |
|   Total Sugars 0g | | |
|     Includes 0g Added Sugars | | **0%** |
| **Protein** 0g | | **0%** |

Not a significant source of cholesterol, vitamin D, calcium, iron, and potassium

*The % Daily Value (DV) tells you how much a nutrient in a serving of food contributes to a daily diet. 2,000 calories a day is used for general nutrition advice.

**On a Nutrition Facts label on a packaged food, calories may be what's written in the biggest type, but it's not what will tell you whether the food is going to cause a spike or not.**

When assessing dry foods, such as cookies, pasta, bread, cereal, cereal bars, crackers and crisps, head to the Total Carbohydrate section. The grams next to Total Carbohydrate and Total Sugars represent the molecules that cause a glucose spike: starches and sugars. The more grams of these, the more the food will lead to a rise in your glucose, fructose and insulin levels and set off the chain reaction that keeps you craving sweet things.

This section also contains the Dietary Fibre line, and as I've described throughout this book, fibre is the only carbohydrate that our body doesn't break down – the more fibre in the food, the flatter the glucose curve after eating it. So here's a tip: for dry foods, look at the *ratio of Total Carbohydrate to Dietary Fibre.*

Select items whose ingredients get the closest to 1 gram of Dietary Fibre for each 5 grams of Total Carbohydrate. Here's how to do it: find the number next to Total Carbohydrate and divide it by five. Try to find a food that has that amount of Dietary Fibre (or as close to it as possible).

Why five? It's an arbitrary cut-off, but I use it because it's close to the ratio we find in fruit such as berries. The science is not exact, but I have found that the closer the food is to this ratio, the flatter the curve it will cause.

Let's say you need to buy bread. Go to the grocery store with your shopping list. Compare options to find items that will keep your spikes low. Put down any loaves that list sugar in their top five ingredients, and of the others choose the one that has the most Dietary Fibre per gram of Total Carbohydrate. *Voilà!*

| Nutrition Facts | | Nutrition Facts | |
|---|---|---|---|
| 15 servings per container | | 15 servings per container | |
| **Serving size** | **30g** | **Serving size** | **29g** |
| Amount per serving | | Amount per serving | |
| **Calories** | **60** | **Calories** | **100** |
| | % Daily Value* | | % Daily Value* |
| **Total Fat** 1g | **1%** | **Total Fat** 0g | **0%** |
| Saturated Fat 0g | **0%** | Saturated Fat 0g | **0%** |
| *Trans* Fat 0g | | *Trans* Fat 0g | |
| **Cholesterol** 0mg | **0%** | **Cholesterol** 0mg | **0%** |
| **Sodium** 110mg | **4%** | **Sodium** 190mg | **8%** |
| **Total Carbohydrate** 25g | **8%** | **Total Carbohydrate** 25g | **8%** |
| Dietary Fiber 14g | **57%** | Dietary Fiber 2g | **8%** |
| Total Sugars 0g | | Total Sugars 7g | |
| Includes 0g Added Sugars | **0%** | Includes 7g Added Sugars | |
| **Protein** 2g | | **Protein** 2g | |
| Vitamin D 2mcg | 10% | Vitamin D | 20% |
| Calcium 260mg | 20% | Calcium | |
| Iron 8mg | 45% | Iron | 30% |
| Potassium 240mg | 6% | Potassium | 2% |
| * The % Daily Value (DV) tells you how much a nutrient in a serving of food contributes to a daily diet. 2,000 calories a day is used for general nutrition advice. | | * The % Daily Value (DV) tells you how much a nutrient in a serving of food contributes to a daily diet. 2,000 calories a day is used for general nutrition advice. | |

**Compare these two cereal labels: Fibre One on the left, Special K on the right. The one on the left has a better fibre-to-carb ratio (14 grams of fibre per 25 grams of total carbs versus 2 grams per 25 grams). The one on the left is a better choice.**

> TRY THIS: Grab something in your pantry that you eat often. Turn to the back of the box and check whether it will cause a spike. Is sugar in the top five ingredients? Is there at least 1 gram of fibre for each 5 grams of total carbohydrate?

### Can I combine these foods with protein and fibre from a different source?

Yes, absolutely you can. You can buy a food that could cause a spike, and then, when you eat it, combine it with fibre, protein and fat – like Oreos with Greek yoghurt and nuts. But you'll make it easier on yourself if you start with ingredients that will help keep your glucose levels steady anyway.

### Should I never buy anything that spikes me or that has sugar in the top three ingredients?

No, no, that would be draconian! The most important thing is to be aware of what spikes you and what doesn't. When I buy ice cream, I'm buying a food that has a ton of sugar in it. It will definitely cause a glucose spike. I know that. It's a conscious decision. I eat it on occasion, rather than every day. For things such as yoghurt, cereal and bread, which I do eat every day, I buy the versions I know will keep my glucose levels steady.

### Beware of the lies

Here is some of the most fun detective work to do – just because a package says something cool on the front doesn't mean that it is good for you. Fancy marketing claims and packaging are just trying to get you to buy their products. For instance, gluten-free, vegan and organic do not mean that the food won't spike you.

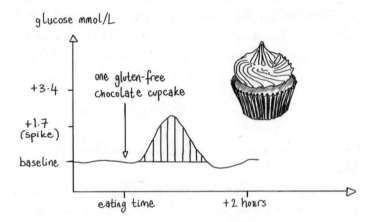

**'Gluten-free' doesn't mean 'healthy'. It just means the food wasn't made with wheat. It can still contain other starches and loads of sugar.**

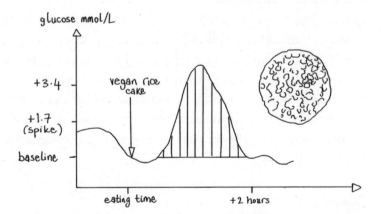

**'Vegan doesn't mean 'healthy'. It just means the food contains no animal products. Same as gluten-free things, these can contain a bunch of starch and sugar.**

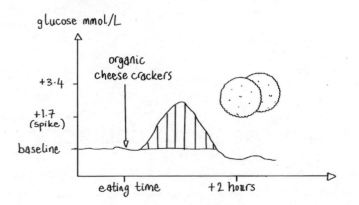

**'Organic' doesn't mean 'healthy'. The food can still contain loads of starch and sugar.**

---

TRY THIS: When you're shopping, stay in the outside aisles of a supermarket. If you shop on the outskirts, you'll find fruit, veggies, dairy, meat, fish – all minimally processed foods. If you venture into the aisles, make sure to use the techniques in this chapter to choose processed foods well. Pretty soon, your brain will become a spike-scanning machine.

---

And one last tip: never go shopping hungry… it messes with your brain. When I do, all vegetables look highly unappetising, and every chocolatey item on the shelves calls my name.

# A DAY IN THE LIFE OF A
# GLUCOSE GOD OR GODDESS

Using the hacks in this book, there are many ways to live like a Glucose God or Goddess. Here is an example from my own life, where I use them to flatten my glucose curves.

**Breakfast:** I had coffee with a splash of whole, not skimmed, milk; the higher fat content helped keep my glucose steady. Two eggs scrambled in a pan with butter and sea salt, served with a couple of tablespoons of hummus on the side. Then a slice of toasted dark rye bread with butter. Before going out the door, I grabbed a square of 80 per cent dark chocolate – I wanted something sweet, and it's best to eat it at the end of a meal rather than on its own at 11 a.m., as I used to do.

Hacks I used:

- Hack 4: Flatten your breakfast curve.
- Hack 6: Pick dessert over a sweet snack.

**At work:** I had black tea (I usually have green tea, but we were out of it).

**Lunch:** I microwaved leftovers: green beans, tahini-baked cod and wild rice, which I ate in that exact order.

The hack I used:

- Hack 1: Eat foods in the right order.

**Afternoon:** While on a walk, I came across the best-looking cookie in the world. So I pulled out a tool from my toolbox: I bought the cookie but didn't eat it just then. I got back to the office, drank a glass of water with a tablespoon of apple cider vinegar swirled in, then five almonds, *then* ate the cookie. About 20 minutes later, it was time to work my muscles to help flatten the curve. So I headed to the bathroom, and there I did 30 squats and 10 push-ups against the sink.

Hacks I used:

- Hack 7: Reach for vinegar before you eat.
- Hack 10: Put some clothes on your carbs.
- Hack 8: After you eat, move.

**Dinner:** I had friends over for dinner. I poured white wine, which contains less glucose and fructose than, say, gin and tonics. I served crudités – raw carrots and sliced hearts of palm – as an appetiser. Once we sat down, I brought out my favourite ham salad and rosemary oven-roasted potatoes on the side. My friends know by now to eat the salad first, then the potatoes, to flatten their glucose curves.

Dessert was strawberries and clotted cream. Twenty minutes after finishing dessert, I got everybody up and we went for a 10-minute walk to the local square. When we got back, my guests were so energised that they all wanted to help with the dishes!

Hacks I used:

- Hack 2: Add a green starter to your meals
- Hack 1: Eat foods in the right order.
- Hack 10: Put some clothes on your carbs.
- Hack 8: After you eat, move.

# YOU ARE SPECIAL

The hacks in this book will work for all of us. No matter who you are, eating your carbs last and adding a green starter to your meal will always flatten your glucose curve. A savoury breakfast is the way to go. Vinegar and exercise will enable you to have your cake and be healthy, too.

However, within a specific category of food – say, dessert – the best option for one person may differ from the best option for her neighbour, as you will see below.

In 2019, I helped my friend Luna get fitted with a glucose monitor and recruited her for a very challenging experiment. First, we ate the exact same breakfast and lunch that didn't spike us. Then, in the middle of the afternoon, I baked cookies, took ice cream out of the freezer, and asked her to eat them at the same time as I did.

What happened was mindblowing.

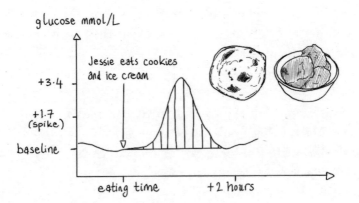

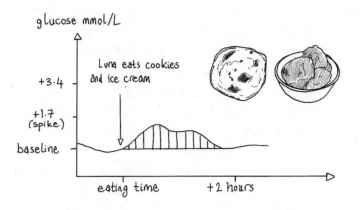

**Two people can have different glucose responses to the same food.**

That's right: a humongous spike for me, barely a spike for her. Neither of us exercised for two hours before or two hours after eating, and no vinegar was consumed. You may be wondering what the heck was going on. Why did the cookies and ice cream shoot my glucose levels through the roof but not hers?

This wasn't a fluke or an isolated experiment. Starting in 2015, research teams around the world have found the same peculiar result: the same food can create different responses depending on the person.

These differences are due to many factors: the amount of baseline insulin we have, our muscle mass, different gut microbes, being more or less hydrated, being more or less rested, being more or less stressed, whether or not we have just worked out – the list goes on. Some studies even found that if you *think* you're about to eat something sugary, that can cause the food to cause a bigger spike for you than someone else.

But although the peaks of our respective spikes may have been different, the general principle applies: if Luna and I had eaten nuts before our cookie and ice cream, both our spikes would have been proportionally smaller.

Individual differences become useful when we look at categories of food. For example, if we look at cookies, that particular food wasn't a good choice for me, while it was probably fine for Luna. So if I'm craving something sweet, I know that cookies aren't the best option for me, whereas apple pie I can handle quite well.

Again, this is rather incomplete. Luna might have had a small spike because she had more insulin in her body – in which case this could actually point to her being less metabolically healthy than me. Science still has a way to go here.

The hacks in this book work for everyone – you don't need to wear a continuous glucose monitor to use them. But if you someday wear one, you may find specific foods that work well for you.

To go even further, you may combine the data from a continuous glucose monitor to an analysis of your gut microbiome, and to your blood fat response to foods. Tim Spector, who provided a blurb for this book, is a scientist who started a company named Zoe that does just that. I've tested the product – and it's clear to me that it's what the future is made of.

# THE END

I am humbled and lucky to hear from many of you every day, and in your messages, there is one resounding conclusion: regardless of what your diet is, your lifestyle, your age, where you live, your past struggles with health, applying the hacks has made a big difference in your lives. As I finish this book and write these words at home in Paris, I want to thank you for giving me the opportunity to share this science.

I know how hard it can be, trying to keep yourself well. Many of us have felt misguided, with conflicting messages coming from all corners. For a long time, I did, too. Indeed, there are many problems with the food advice we get these days, not the least of which is that it is rarely totally impartial.

Maybe because of this you've been following one health regime or another that has not only not worked but has made your condition worse. Maybe your body has felt like a black box. Maybe you've felt tired for years; maybe you're battling with cravings or weight gain or a chronic condition. Maybe you are depressed, have fertility troubles, or are inching closer and closer to type 2 diabetes. Maybe you're lost as to how to manage your type 1 diabetes or your gestational diabetes. Maybe you take medication for a condition you were told you can't do anything about.

I hope you have learned from reading this book that the symptoms you are having are actually powerful messages. Your body is speaking to you.

My aim has been to bring up-to-date, objective science into the realm of action, to turn unbiased research into realistic tools, to arm you with knowledge about how your body works, and to help you feel amazing.

What are you going to do? Are you going to listen to your body, understand the glucose lever in the cockpit and get yourself back to cruising altitude? I hope you are. As you do, remember, it's important to be kind to yourself in the process. I hope you will then go on to help your parents, siblings, children, friends and acquaintances do the same. Together we can help everyone reconnect with their bodies, one person at a time. I hope you'll let me know how it goes. I'd love to hear about your journey. Reach out to me on Instagram at @glucosegoddess.

# RECIPES

## BREAKFASTS

### Brilliant berry and almond butter protein smoothie

I first discovered morning smoothies in Dr Mark Hyman's book *The Blood Sugar Solution*, a great inspiration of mine. Over the years I have refined the recipe. The cacao nibs are my favourite addition. Sometimes I want a thicker smoothie, so I reduce the water or milk to 50ml.

*Serves 1*
½ small avocado (roughly 50g)
¼ banana (roughly 25g)
40g frozen mixed berries
3 tbsp protein powder
1 tbsp almond butter
1 tsp groundnut oil
150ml full-fat milk, unsweetened nut milk or filtered water
1 tsp cacao nibs

1. Place all the ingredients, except the cacao nibs, in a blender and blitz until smooth. Serve immediately, topped with the nibs.

# Creamy steel-cut oats with nut butter, berries and cacao nibs

I get a lot of queries about oats from the community. Keep in mind that oats are starches, so they naturally tend to spike our glucose levels. The trick here is to combine them with ingredients that will flatten the curve and also make them taste delicious. Try to avoid honey, maple syrup or sugar and add sweetness with berries instead.

*Serves 2*
250ml full-fat milk or unsweetened nut milk
1 tsp ground cinnamon
60g steel-cut oats
2 tbsp almond butter
2 tbsp full-fat Greek yoghurt
60g mixed berries, such as blueberries, raspberries and strawberries
1 tsp cacao nibs

1. Pour the milk into a medium-sized saucepan, along with the cinnamon, 500ml water and a pinch of salt. Bring to the boil, then reduce to a simmer and stir in the oats.

2. Cook for 20-25 minutes, stirring frequently.

3. Divide the porridge between two bowls and swirl the nut butter on top, followed by the Greek yoghurt, berries and cacao nibs.

# Green shakshuka

This is a really impressive dish to serve to your friends for Sunday brunch. It's always met with 'ooohs' and 'ahhhs' when I set it down on the table.

*Serves 4*
1 medium leek, sliced
1 medium fennel bulb, roughly chopped
1 garlic clove
2 tbsp olive or avocado oil
100g baby spinach
10g dill, roughly chopped (reserve some for garnish)
2 tbsp tahini (see note below)
½ tsp chilli flakes
4 medium free-range eggs

1. Preheat the oven to 200°C/fan 180°C/gas mark 6. Heat the oil in a medium-sized non-stick ovenproof frying pan and sauté the leek, fennel and garlic until softened and just starting to take on some colour – about 7 minutes.

2. Add the spinach and allow it to wilt, then stir in the dill, tahini and chilli flakes with some seasoning.

3. Make four holes in the mixture with a large spoon and crack in the eggs. Allow them to cook for a couple of minutes on the hob and then transfer the pan to the oven for 5-7 minutes, or until the eggs are just set.

4. Garnish with some dill before serving.

Tip – If the tahini you are using is not of a pouring consistency, simply mix with a few tablespoons of boiling hot water.

# Green Goddess soufflé omelette with rocket, avocado and smoked salmon

I've been trying to perfect my omelette game for a while. This one is easy to make, and an excellent way to use up leftovers.

*Serves 1*
2 eggs, separated
2 spring onions, finely chopped
30g rocket, finely chopped
5g dill, finely chopped
40g feta cheese, crumbled
Knob of butter
1 slice of smoked salmon
Few slices of cucumber
½ small avocado
1 tbsp sour cream

1. Preheat the grill.

2. Whisk the egg whites into stiff peaks. Gently fold in the spring onions, rocket, dill and feta cheese. Add the egg yolks and some seasoning, then mix everything together well, trying not to deflate the egg whites too much.

3. Melt the butter in a non-stick frying pan, and when sizzling, pour in the egg mixture, tipping the pan so that it covers the base. Cook for 3 minutes on the hob, then place it under the grill for another 1-2 minutes, by which time the eggs should be set and golden.

4. Transfer the omelette to a plate, and place the salmon, cucumber, avocado and sour cream on one half of it. Season, flip the other half over and serve.

# Pumpernickel avocado toast with pickled radish

Not all breads are created equal as far as our glucose levels are concerned. In general, white bread will make them shoot up, then crash, leaving us feeling hungry again. Pumpernickel bread, on the other hand, is packed with fibre. The result is a steady curve, less inflammation in our body and more goodness. The pickled radish takes seconds to make and adds a nice kick to the creamy cheese and avocado combo.

*Serves 2*
5 pink radishes
1 tsp fennel seeds, crushed
3 tsp red wine vinegar
½ tsp sea salt
60g cream cheese
2 slices of pumpernickel bread, toasted
1 avocado, sliced and dressed with a little lemon juice
  (to stop it from browning)
Fresh chopped dill, to garnish (optional)
Avocado oil, to garnish (optional)

1. First make the pickle. Roughly grate the radishes into a small bowl and add the crushed fennel seeds, vinegar and salt. Stir well and set aside while you prepare the rest of the dish.

2. Spread the cream cheese over the bread and place the avocado on top. Squeeze the excess liquid out of the radishes, then scatter them over the avocado.
3. Garnish with fresh dill and a drizzle of avocado oil.

Tip – Smoked salmon makes a great addition to this dish.

# 'GREEN' STARTERS

## Pan-fried courgette with avocado oil and Parmesan

I've been making this dish for as long as I can remember –
it's a real comfort food for me. I think it's the best way to eat
courgettes – browning them really enhances their flavour.
Some chopped fresh chilli or herbs would add a splash of
colour and another layer of taste.

*Serves 2*
2 medium courgettes (roughly 500g), sliced into thin discs
2 tbsp avocado oil
50g Parmesan, finely grated


1. Fry the courgettes in the avocado oil for about 8 minutes
or until they begin to take on some colour and have
softened slightly. Season well with sea salt and ground
black pepper. Transfer to a serving dish and scatter over the
grated Parmesan.

# Simple green salad with artichoke, hazelnuts and feta

My go-to salad. I use vinegar in the dressing to flatten the glucose curve of any starch I eat afterwards. The mixed fresh herbs provide a big punch of flavour.

*Serves 4 as a side, 2 as a main*
200g mixed salad leaves
Large handful of fresh herbs, such as parsley, chives, oregano, basil or mint
70g artichoke hearts (from a jar), drained and sliced
50g blanched hazelnuts, toasted
100g feta cheese, crumbled

*Dressing*
2 spring onions, finely sliced
1½ tbsp live apple cider vinegar
3 tbsp avocado or olive oil
1 tsp Dijon mustard

1. Start by making the dressing. Soak the spring onions in the cider vinegar for 5 minutes, then add the avocado oil, mustard and some seasoning and whisk to emulsify.

2. Place the salad ingredients in a bowl, pour over the dressing and toss everything together. Serve immediately.

# Tangy red cabbage with pomegranate seeds and coriander

After learning this recipe, I finally understood what red cabbage was for! This little salad has become a family staple – easy, fresh and as vibrant in flavour as it is in appearance.

*Serves 4 as a side, 2 as a main salad*
½ red cabbage, finely sliced
2 tbsp live apple cider vinegar
Juice of ½ orange
60g pomegranate seeds
20g fresh coriander, roughly chopped

1. Toss the cabbage in the vinegar in a bowl, then add the remaining ingredients, season and serve immediately.

# Broad bean, chicory and gorgonzola soup

This soup is deliciously delicate, the tender broad beans and crunchy chicory enveloped in a silky smooth broth.

*Serves 2*
1 onion, finely chopped
2 garlic cloves, finely chopped
Leaves from 3 sprigs of thyme
2 tbsp olive oil
700ml vegetable or chicken stock
50g gorgonzola cheese (dolcelatte will also work)
1 head of chicory, finely sliced
400g fresh broad or fava beans, shelled (200g shelled weight)
Squeeze of lemon juice

1. In a medium saucepan, sauté the onion, garlic and thyme in the olive oil for 3-4 minutes. Pour in the stock, bring to the boil and simmer for 5 minutes.

2. Add the cheese, then remove the broth from the heat and blitz until smooth.

3. Return the pan to the heat, add the chicory and beans and simmer for a couple of minutes. Follow with the lemon juice and season to taste.

# Tenderstem broccoli with basil, lemon, chilli and Parmesan

This makes a brilliant starter, main or side dish. It's also perfect for breakfast, topped with a fried egg.

*Serves 2 as a main or 4 as a starter/side*
450g tenderstem broccoli
4 tbsp avocado oil
Juice of 1 lemon and zest of ½
30g basil leaves, roughly torn
1 red chilli, deseeded and finely chopped
20g pine nuts, toasted
20g Parmesan, grated

1. Bring a medium saucepan of salted water to the boil and cook the broccoli for 2 minutes.

2. Meanwhile, in a large bowl, mix the avocado oil, lemon juice and zest, basil and chilli. Season well with salt and freshly ground black pepper.

3. When the broccoli is ready, drain it and toss it in the dressing, ensuring that it gets a thorough coating. Scatter over the toasted pine nuts and grated Parmesan before serving.

# MAIN DISHES

## Chicken traybake with baby potatoes, olives, capers and cherry tomatoes

One-pot cooking at its very best. This dish is as suitable for cosy weeknight suppers as it is for show-off entertaining!

*Serves 4*
550g baby potatoes
300g cherry tomatoes
60g pitted black olives, chopped in half
Few sprigs of rosemary and/or oregano (optional)
3 tbsp olive oil
4 chicken legs, skin on, bone in (roughly 1kg)
25g capers (from a jar), drained
20g fresh parsley, roughly chopped

1. Preheat the oven to 200°C/fan 180°C/gas mark 6. Place the potatoes, tomatoes, olives and herbs in a medium-sized roasting pan. Sprinkle over some seasoning, then toss everything in 2 tbsp of the olive oil.

2. Place the chicken legs on top, drizzle the remaining olive oil over the skin, season again and roast, uncovered, for 1 hour.

3. Remove the pan from the oven and stir in the capers. Roast for a further 15 minutes. Divide the chicken and veg between 4 plates, making sure to spoon out all the cooking juices. Garnish with lots of fresh parsley before serving.

# Charred aubergines and tomatoes with chickpeas, capers, and oregano and basil ricotta

The aubergines that sit atop this dish become beautifully charred in a very hot oven, while the tomatoes, capers, oregano and chickpeas cook and melt together beneath to create a beautiful base.

*Serves 4*
500g cherry tomatoes
2 tbsp tomato purée
50g sundried tomatoes (in oil), roughly chopped
2 x 400g tins of chickpeas, drained
30g capers (from a jar), drained
Leaves from 4 sprigs of oregano (or 1 tsp dried oregano)
3 small aubergines, cut into 6 wedges lengthways
2 tbsp olive oil
100g ricotta
Leaves from 2 sprigs of basil, finely chopped

1. Preheat the oven to 240°C/fan 220°C/gas mark 9. Place the tomatoes, tomato purée, sundried tomatoes, chickpeas, capers and oregano in a medium-sized roasting pan. Pour in 250ml water and season with sea salt and freshly ground black pepper. Mix everything together well.

2. Arrange the aubergine slices neatly on top and drizzle with olive oil. Season once more and put the dish in the oven for 30-40 minutes, or until the aubergines are nicely charred.

3. Mix the basil and ricotta together, and put a dollop on top of each serving.

# Coconut and lentil curry with spinach

There's something so satisfying about stirring spinach leaves into a hot dish – they cook in seconds while maintaining their vibrant greenness. If you're cooking for four, double the quantities and increase the cooking time slightly. Any leftovers can be stored in a jar in the fridge for a couple of days.

*Serves 2*
1½ tbsp olive or coconut oil
1 onion, finely chopped
3 garlic cloves, finely chopped
50g root ginger, peeled and finely chopped
½ tsp ground coriander
1 tsp ground cumin
1 tsp ground turmeric
120g green lentils
400g tin of coconut milk
1 chicken or vegetable stock cube
50g spinach
20g fresh coriander, roughly chopped
Juice of ½ lime

1. Heat the oil in a medium saucepan and fry the onion, garlic and ginger for 3-4 minutes. Add the spices and stir for 30 seconds to release their aromas.

2. Stir in the lentils, coconut milk, crumbled stock cube and 400ml water. Simmer, with the lid on, for 25-30 minutes, or until the lentils are cooked, but retain a little bite.

3. Add the spinach, coriander and lime juice and stir to combine. Serve with a spoonful of basmati rice, or cauliflower rice (you can buy this readymade).

# Cod with tahini, pine nuts and spinach

I confess that I was seduced by the man who made this dish for me. It's sexiness on a plate.

*Serves 2*
2 skinless cod fillets (roughly 250g)
2 tbsp olive or avocado oil, plus extra for drizzling
2 shallots, finely chopped
Leaves from 2-3 sprigs of thyme
1 tsp fennel seeds, crushed
40ml tahini (loosened with a little hot water if required)
10g fresh dill, finely chopped, plus extra to garnish
Juice of ½ lemon
200g spinach
20g pine nuts, toasted

1. Preheat the oven to 220°C/200°C fan/gas mark 7. Place the cod fillets on a baking tray lined with baking paper, drizzle with a little oil, season and roast in the oven for 10 minutes.

2. Meanwhile, make the sauce. Sauté the shallots in the oil with the thyme and fennel seeds for about 5 minutes, or until softened. Add the tahini, along with 80ml boiling water, and stir, over a very low heat, until the sauce is nice and smooth. Add more hot water if necessary. Stir in the dill and lemon juice with some salt and freshly ground black pepper.

3. In a separate pan, wilt the spinach with a splash of boiling water, then add some seasoning.

4. Place the cod on top of the spinach, spoon over the tahini sauce and garnish with the pine nuts and some fresh dill. Serve with a pile of steamed green veg.

# Greens and grains salad with chicken, pomegranate, pistachios and a lemon yoghurt dressing

This wholesome and filling salad will get your tastebuds popping with its wonderful combination of textures and vibrant lemony zing! To reduce preparation time, you could use ready-cooked grains and chicken.

*Serves 2*
2 small chicken breasts (roughly 350g), sliced in half lengthways
1 tbsp olive oil
150g quinoa (mix of white, red and black if possible)
250g tenderstem broccoli, chopped into bite-size pieces
100g asparagus, woody ends trimmed, chopped into bite-size pieces
100g sugar snap peas, cut in half
50g pomegranate seeds
25g pistachios, roughly chopped
10g fresh coriander leaves (optional)

*Dressing*
3 tbsp (60g) full-fat Greek yoghurt
1½ tbsp olive oil
Zest of 1 lemon and juice of ½
Sea salt and ground black pepper

1. Preheat the oven to 200°C/180°C fan/gas mark 6. Place the chicken in a roasting pan, drizzle with olive oil, season and roast for 20-25 minutes. When ready, cut into small chunks.

2. Meanwhile, place the quinoa in a small saucepan, cover with water, add a pinch of salt and simmer for 12-15

minutes. When ready, drain and keep warm.

3. In another saucepan, cover the broccoli, asparagus and sugar snap peas with water, add a generous pinch of salt, bring to the boil and simmer for 1 minute. Drain and keep warm.

4. Mix all the dressing ingredients together in a bowl.

5. Put the quinoa, vegetables and chicken in a large bowl and toss them in the dressing. Garnish with pomegranate seeds, pistachios, coriander (if using) and serve immediately.

# Jessie's grandmother's pot-baked white cabbage with thinly sliced ribeye steak

My Brazilian grandmother once served this cabbage dish for lunch and I was instantly hooked. I made it every week for two months after that. It's deceptively simple and incredibly good for your glucose levels. I promise it will become a staple for you, too.

*Serves 4*
80g butter
3 large garlic cloves, roughly chopped
1 small white cabbage (roughly 850g), thinly sliced
120g goat's cheese, rind removed and roughly chopped
50g pine nuts, toasted
60g pomegranate seeds
20g parsley, roughly chopped
2 large ribeye steaks, 2-3cm thick, at room temperature
Olive oil

1. Melt the butter in a medium-sized heavy-based saucepan (with a lid) and sauté the garlic for 2-3 minutes. Add the cabbage, along with 80ml water. Give it a good stir, then put the lid on and let it cook for 15-20 minutes. Every 5 minutes or so, take the lid off and stir again. The cabbage will reduce in size and become translucent, tender and silky.

2. Now add the goat's cheese and stir until melted. Remove from the heat and tip in the pine nuts, pomegranate seeds and parsley. Season well with sea salt and freshly ground black pepper. Put the lid back on to keep the cabbage warm while you cook the steaks.

3. Place a griddle pan over a very high heat. Coat the steaks

with olive oil, some salt and lots of pepper. When the pan is almost smoking hot, griddle the steaks for 2-3 minutes on each side (you may need a little longer for a thicker steak or if you don't like your steak too rare). When ready, rest the meat, covered, for 5-6 minutes, then slice it thinly and serve it with the cabbage.

Notes on cooking steak

– A smoking-hot pan is absolutely essential (or use a barbeque).

– Make sure you use a pan that can accommodate both steaks comfortably.

– Oil your steak rather than the pan.

– Leave the steaks to cook for the full 2-3 minutes before turning.

# Linguine with oven-baked feta, oregano and tomatoes

My flatmate at uni in London taught me this recipe. She's Italian, so you know it's legit.

*Serves 4*
500g cherry tomatoes
200g feta cheese
Leaves from 4 sprigs of oregano (or 1 tsp dried oregano)
3 tbsp olive oil
350g linguine
70g rocket or spinach, roughly chopped
Squeeze of lemon juice

1. Preheat the oven to 200°C/fan 180°C/gas mark 6. Place the tomatoes, feta cheese, oregano and olive oil in a medium/large roasting pan. Season and bake for 30 minutes.

2. About 10 minutes before the end of cooking time, put a pan of salted boiling water on the hob and cook the linguine until al dente, around 8 minutes. Drain, reserving some of the cooking liquid.

3. Remove the roasting pan from the oven, add a splash of the pasta cooking water and use the back of a fork to mash everything together. Stir in the rocket (or spinach) and lemon juice.

4. Tip the pasta into the pan, toss everything together until the pasta is thoroughly coated and serve immediately.

# Spicy pork tacos with black bean, tomato and lime salsa, avocado and baby gem

This is a wonderful dish to share with guests. Put everything out in bowls and let everyone assemble their own. You can replace the tacos with large salad leaves if you prefer.

*Serves 4*
4 tbsp soy sauce
1 fresh chilli, deseeded and finely chopped
Juice of 2 limes
500g pork fillet, sliced as thinly as possible
400g tin of black beans, drained
100g tomatoes, quartered
20g fresh coriander, roughly chopped
1 small baby gem lettuce, finely sliced
1 avocado, sliced and dressed with a little lime juice
8 small soft flour tortilla wraps (if you can only find large wraps, you will need just one per person)
120ml sour cream

1. Begin by cooking the pork. Place the soy sauce, chilli and the juice of 1 lime in a medium-sized saucepan. Simmer for about a minute, until the mixture has reduced and thickened slightly. Add the pork and stir-fry for just a couple of minutes – long enough to cook through but not so long that it becomes tough. Set it aside and keep it warm.

2. Mix the black beans, tomatoes, coriander and the remaining lime juice together in a serving bowl with some seasoning and set aside.

3. Place the avocado, baby gem and sour cream in three separate serving bowls.

4. When you're ready to eat, toast each tortilla wrap, either over a toaster or in the oven. Pile them up on a plate as you go, keeping them warm by covering them with a tea towel.

5. When ready, get everyone to the table to assemble their own taco.

# Super-quick clam and chorizo bean stew

A rich spicy stew, full of Mediterranean flavours.

*Serves 4*
2 tbsp olive oil
1 medium onion, finely chopped
2 garlic cloves, finely chopped
1 red chilli, deseeded and finely chopped
100g spicy chorizo, cut into 1cm cubes
1 tbsp tomato purée
1 tsp paprika
400g cherry tomatoes
120ml white wine
2 x 400g tins of borlotti beans, drained
1 chicken or vegetable stock cube
700g clams
30g fresh parsley, roughly chopped
1 lemon, cut into 4 wedges

1. Heat the oil in a wide, medium-sized casserole dish and sauté the onion, garlic and chilli for 3-4 minutes, or until softened. Add the chorizo and fry for 3 more minutes.

2. Now add the tomato purée, paprika, tomatoes, wine, beans and stock cube. Stir to combine and pour in 500ml water. Bring to the boil, then reduce to a simmer for 20 minutes. As the tomatoes burst and rise to the surface, gently mash them with the back of a wooden spoon.

3. Add the clams to the pan and cover with a lid. Cook for 5 minutes, shaking the pan every now and then. By this time, all the shells should have opened (discard any that haven't). Check for seasoning, garnish with parsley and serve with wedges of lemon.

# Stir-fried cauliflower rice bowl with tempeh

This nutritious stir-fry is great for a weeknight supper. For an added hit of protein, stir through a beaten egg just before serving. To cut the cooking time, blitz the spring onion, ginger and garlic in a food processor.

*Serves 2*
100g tempeh, cut into 1cm cubes
2 tbsp olive oil
100g spring onions, finely chopped
50g root ginger, peeled and finely chopped
2 garlic cloves, finely chopped
1 red pepper, deseeded and diced
100g baby corn, halved and diced
50g kale, thick stalks removed and chopped
200g cauliflower rice (shop bought)
25g fresh coriander, roughly chopped
2 tbsp soy sauce
Juice of 1 lime
25g cashew nuts, roughly chopped (optional)

1. In a wok or large frying pan, stir-fry the tempeh in 1 tbsp olive oil until golden brown, then set it aside. Add the remaining oil to the pan and stir-fry the spring onions, ginger and garlic for 2-3 minutes.

2. Return the tempeh to the pan, along with the red pepper, baby corn and kale. Add 1-2 tbsp water and cook, stirring, for about 3 minutes, or until the kale has wilted.

3. Now stir in the cauliflower rice and give it a minute or so to heat through. Remove the pan from the heat and add the coriander, soy sauce and lime juice. Scatter over the cashew nuts (if using) and serve immediately.

# DESSERTS

## Baked rhubarb posset with almonds

I love rhubarb and used to grow it outside my window in east London. The leaves got so huge they were bigger than my face! Grated chocolate would also make a nice garnish for this scrumptious pud.

*Serves 2*
2 sticks of rhubarb (roughly 300g), chopped into 2cm pieces
Zest and juice of 1 orange
40g soft brown sugar
240g full-fat Greek yoghurt
20g flaked almonds, toasted

1. Preheat the oven to 200°C/fan 180°C/gas mark 6. Gently toss the rhubarb with the orange zest and juice and sugar in a small roasting pan and bake in the oven for 30 minutes. Remove from the oven and allow to cool.

2. Mix the rhubarb with the yoghurt and serve topped with toasted almonds.

# Chocolate and banana semifreddo

Here I am again with my love of chocolate. The eggs in this recipe add protein and the double cream adds fat – elements that combine to create a dessert that keeps our glucose steady. You'll also notice there isn't much sugar in here. So win win.

*Serves 6-8*
1 ripe banana
2 tbsp unsweetened cocoa powder
70g caster sugar
4 medium free-range eggs, separated
300ml double cream

1. Line a 1-litre loaf tin with clingfilm. Line up three bowls and an electric whisk.

2. Place the egg whites in one bowl and the double cream in another. In the third bowl, mash the banana, add the cocoa powder and sugar and mix until thoroughly combined, before stirring in the egg yolks.

3. Now, whisk the egg whites until they form stiff peaks. Finally, whisk the double cream until it holds its shape. If you do it in this order, you don't need to wash the whisk in between.

4. Fold the cream into the banana mixture, then gently fold in the egg whites. Transfer the mixture to the lined baking tin and place it in the freezer.

5. Remove the semifreddo from the freezer 20 minutes before serving so that it softens slightly.

# Eton mess with cacao nibs

A delicious dessert that makes the most of juicy summer berries. If you find yourself short of time, replace the cream with Greek yoghurt – that way you won't need to do any whisking.

*Serves 4*
225ml double cream
½ tbsp icing sugar
4 shop-bought meringues
225g strawberries or raspberries (or a mix of both)
2 tbsp cacao nibs

1. Using an electric or balloon whisk, whisk the cream with the icing sugar until it is just holding its shape.

2. In another bowl, smash the meringues into chunky pieces. Add the cream and stir gently to combine. Finally, fold in the berries and serve immediately, topped with cacao nibs.

# Mixed berry crumble with nutty topping

This crunchy nutty topping provides a nutritious alternative to traditional flour-based versions. Flour turns to glucose when digested but nuts do not. Use any mix of berries for this recipe and feel free to experiment with different nut butters.

*Serves 4*
150g ground almonds
100g caster sugar
40g butter, cut into cubes
30g peanut butter (or any other nut butter)
50g hazelnuts, roughly chopped
100g almonds, roughly chopped
550g frozen berries
1 tbsp cornflour

1. Preheat the oven to 200°C/fan 180°C/gas mark 6. Mix the ground almonds and 40g of the caster sugar together in a bowl. Add the butter, and using your fingertips, rub in the ground almonds until everything is well combined. Stir in the peanut butter, hazelnuts and almonds.

2. Place the berries in a medium-sized baking dish and stir in the remaining sugar and cornflour. Scatter over the crumble mix, then place in the oven and bake for 25-30 minutes, or until the berries are bubbling and the crumble is golden brown. Serve with some Greek yoghurt or double cream.

# White chocolate, melon and mint

A magically light and refreshing dessert for a summer's day.

*Serves 4*
100g caster sugar
900g cantaloupe melon (or use a mixture of varieties)
Generous handful of fresh mint
200g white chocolate, roughly chopped

1. Place the sugar in a medium saucepan with 100ml water and 3 sprigs of mint. Bring to a gentle simmer and stir from time to time until the sugar has completely dissolved. Set aside to cool and remove the mint.

2. Using a melon baller, scoop the flesh out of the melon. Alternatively, slice it into 2cm chunks.

3. Arrange the melon on a large serving platter, pour over the syrup, scatter with mint leaves and garnish with the white chocolate.

# Gooey chocolate brownie with smashed raspberry Greek yoghurt

You might wonder why I add Greek yoghurt to my brownie. It's because I never let my carbs walk around naked...

*Makes 12 brownies*
180g dark chocolate (minimum 70 per cent solids)
180g unsalted butter
300g caster sugar
130g plain flour
½ tsp baking powder
3 medium free-range eggs
100g full-fat Greek yoghurt
50g raspberries

1. Preheat the oven to 180°C/fan 160°C/gas mark 4. Line a baking tray (30cm x 20cm) with baking paper. Place the chocolate and butter in a heatproof bowl and set over a saucepan of simmering water (be sure not to let the base of the bowl touch the water).

2. When the chocolate and butter have completely melted, remove from the heat and stir in the sugar, followed by the flour and baking powder. Crack in the eggs and beat the mixture until completely smooth.

3. Pour into the prepared tin and bake for 20 minutes. Remove from the oven and allow to cool before cutting into 12 squares.

4. Mash the raspberries with the back of a fork and marble through the yoghurt. Serve a dollop on each brownie.

Tip – For some added crunch, scatter pistachios over the brownie before baking.

# ACKNOWLEDGEMENTS

This book took a village. And what a village! I'd like to thank the people in the Glucose Goddess community who contributed their glucose data, their stories and their passion to this work. This book was born out of the movement that we're building together.

I'd like to thank Susanna Lea, agent of my dreams, for bringing her experience, humour and wisdom to my life. Thank you to Mark Kessler and everyone at SLA for welcoming me. Thank you to the team at Simon & Schuster and to Emily Graff for your enthusiasm and commitment. Thank you to Short Books, to Rebecca Nicolson and Aurea Carpenter for your force and dedication. Thank you, Evie Dunne, for your brilliant illustrations.

Thank you to Robert Lustig for the feedback that I terribly needed. Thank you to Elissa Burnside, my first friend and my first reader, for your spirit and love. Thank you to Franklin Servan-Schreiber for channelling the universe for me. Thank you to David Servan-Schreiber for paving the way.

To my friends, thank you for being the best and for sharing this adventure with me. Dario, thank you for your unconditional love. Thank you, Sefora, for helping me through my life. Thank you to Alice, Paul, Ines, Mathieu, Arthur, Jasmyn, and my entire family. Thank you, Dad, for your kindness. Thank you, Mom, for being *my* goddess.

Thank you to Anne Wojcicki, Kevin Ryan and Thomas Sherman for believing in me and guiding my path.

Thank you to all the scientists who ran the studies around

the world, and to those before them, whose shoulders this work rests upon. Thank you to Axel Esselmann and Lauren Kohatsu for believing in this work from the beginning. Thank you to everyone at 23andMe who shaped my understanding of how we can make science accessible. Thank you, Bo, for your help getting this crazy project off the ground.

Closing this book out, I also want to say thank you to myself. Thank you for trusting and following what makes your soul light up. Waking up and going after it. While it wasn't an easy journey, I'm glad the idea picked me – and I hope I did it justice.

# Endnotes

## Dear Reader

*"What we eat affects the 30 trillion cells"*: Ron Sender et al., "Revised estimates for the number of human and bacteria cells in the body." *PLoS Biology* 14, no. 8 (2016): e1002533.

*"Our nutritional choices are influenced by billion-dollar marketing campaigns"*: Rudd Center for Food Policy and Obesity, *Increasing disparities in unhealthy food advertising targeted to Hispanic and Black youth,* January 2019, accessed August 30th, 2021, https://media.ruddcenter.uconn.edu/PDFs/TargetedMarketingReport2019.pdf

*"These are usually justified under"*: Robert Lustig, *Metabolical: The Lure and the Lies of Processed Food, Nutrition, and Modern Medicine* (New York: Harper Wave, 2021).

*"processed foods and sugar are inherently bad"*: Robert Lustig, *Metabolical: The Lure and the Lies of Processed Food, Nutrition, and Modern Medicine* (New York: Harper Wave, 2021).

*"88 per cent of Americans are likely to have dysregulated glucose levels"*: Joana Araújo et al., "Prevalence of optimal metabolic health in American adults: National Health and Nutrition Examination Survey 2009–2016." Metabolic syndrome and related disorders 17, no. 1 (2019): 46-52.

*"too much insulin is one of the main drivers of"*: Benjamin Bikman, *Why We Get Sick: The Hidden Epidemic at the Root of Most Chronic Disease and How to Fight It* (New York: BenBella, 2020).

*"too much fructose increases the likelihood of"*: Robert Lustig, *Metabolical: The Lure and the Lies of Processed Food, Nutrition, and Modern Medicine* (New York: Harper Wave, 2021).

## How I got here

*"your genes can increase your likelihood of developing type 2 diabetes"*: Michael Multhaup at al., *The science behind 23andMe's Type 2 Diabetes report*, 2019, accessed August 30th, 2021, https://permalinks.23andme.com/pdf/23_19-Type2Diabetes_March2019.pdf

*"top athletes started to wear CGMs, too"*: Mark Hearris et al., "Regulation of muscle glycogen metabolism during exercise: implications for endurance performance and training adaptations." *Nutrients* 10, no. 3 (2018): 298.

*"people without diabetes could have highly dysregulated glucose levels"*: Heather Hall et al., "Glucotypes reveal new patterns of glucose dysregulation," *PLoS Biology* 16, no. 7 (2018): e2005143, https://pubmed.ncbi.nlm.nih.gov/30040822/

## PART I: WHAT IS GLUCOSE?

### Chapter 1: Enter the cockpit: why glucose is so important

*"only 12 per cent of Americans are metabolically healthy"*: Joana Araújo et al.,

"Prevalence of optimal metabolic health in American adults: National Health and Nutrition Examination Survey 2009–2016," Metabolic syndrome and related disorders 17, no. 1 (2019): 46-52, https://pubmed.ncbi.nlm.nih.gov/30484738/

*"Waist size is better for predicting underlying disease than BMI is"*: Division of Nutrition, Physical Activity, and Obesity, National Center for Chronic Disease Prevention and Health Promotion, *Assessing your Weight*, CDC, September 17th, 2020, accessed August 30th, 2021, https://www.cdc.gov/healthyweight/assessing/index.html

## Chapter 2: Meet Jerry: how plants create glucose

*"plants make extra glucose during the day"*: Gregory MacNeill et al., "Starch as a source, starch as a sink: the bifunctional role of starch in carbon allocation," *Journal of Experimental Botany* 68, no. 16 (2017): 4433-4453, https://pubmed.ncbi.nlm.nih.gov/28981786/

*"plants also transform some of their glucose into extra-sweet molecules called fructose"*: M D. Oesten et al., Castellion. "Sweetness relative to sucrose (table)." *The World of Chemistry: essentials,* 4th edition. Belmont, Thomson Brooks/Cole 359 (2007).

## Chapter 3: A family affair: how glucose gets into the bloodstream

*"Every second, your body burns"*: The body uses 200 grams of glucose daily. Glucose has a molar mass of 180 g/mole. Per day, the body therefore uses 0.1111 mole of glucose. 1 mole has $6.02214076 \times 1023$ molecules in it. So the body uses uses 6.6912675e+23 molecules of glucose per day. A day has 86400 seconds in it. 7.7445226e+18 molecules per second.

Jeremy M. Berg, *Biochemistry,* 5th edition (New York: W. H. Freeman, 2002), Section 30.2.

*"if each glucose molecule were a grain of sand"*: About 5 sextillion (5 x 10"21 grains of sand on earth)

Jason Marshall, *How Many Grains of Sand Are on Earth's Beaches?*, Quick and Dirty Tips, 2016, accessed August 30th, 2021, https://www.quickanddirtytips.com/education/math/how-many-grains-of-sand-are-on-earth-s-beaches?page=all

*"using the same enzyme that plants use"*: Liangliang Ju et al., "New insights into the origin and evolution of α-amylase genes in green plants," *Scientific reports* 9, no. 1 (2019): 1-12, https://pubmed.ncbi.nlm.nih.gov/30894656/

*"Fructose is a little more complicated"*: Cholsoon Jang et al., "The small intestine converts dietary fructose into glucose and organic acids," *Cell Metabolism* 27, no. 2 (2018): 351-361, https://www.ncbi.nlm.nih.gov/pmc/articles/PMC6032988/#SD1

*"In 1969, a cohort of scientists wrote a"*: IUPAC, Comm, and IUPAC-IUB Comm. "Tentative rules for carbohydrate nomenclature. Part 1, 1969," *Biochemistry* 10, no. 21 (1971): 3983-4004, https://pubs.acs.org/doi/abs/10.1021/bi00797a028

*"some humans evolved in areas without"*: Mindy Weisberger, "Unknown Group

of Ancient Humans Once Lived in Siberia, New Evidence Reveals," *Live Science*, 2019, accessed August 30th, 2021, https://www.livescience. com/65654-dna-ice-age-teeth-siberia.html

*"Scientists know that humans' prehistoric diet"*: Marion Nestle, "Paleolithic diets: a sceptical view," *Nutrition Bulletin* 25.1 (2000): 43-47, https:// onlinelibrary.wiley.com/doi/abs/10.1046/j.1467-3010.2000.00019.x

*"They adapted to the unique food supply"*: Peter Ungar, *Evolution's Bite: A Story of Teeth, Diet, and Human Origins* (Princeton University Press, 2017).

**Chapter 4: Seeking pleasure: why we eat more glucose than before**

*"Fibre-packed seeds"*: U.S. Department of Agriculture, "Wheat bran, crude," FoodData Central, 2019, accessed August 30th, 2019, https://fdc.nal.usda. gov/fdc-app.html#/food-details/169722/nutrients

*"starchy bread"*: U.S. Department of Agriculture, "Bread, white, commercially prepared," FoodData Central, 2019, accessed August 30th, 2019, https:// fdc.nal.usda.gov/fdc-app.html#/food-details/325871/nutrients

*"This is the same chemical that is released"*: Nora Volkow et al., "The brain on drugs: from reward to addiction," *Cell* no. 162.4 (2015): 712-725, https:// pubmed.ncbi.nlm.nih.gov/26276628/

*"In a 2016 study, mice were given a lever"*: Vincent Pascoli et al., "Sufficiency of mesolimbic dopamine neuron stimulation for the progression to addiction," *Neuron* 88, no. 5 (2015): 1054-1066, http://www. addictionscience.unige.ch/files/8214/6037/1136/NeuronVP2015.pdf

*"Ancestral bananas (top image)"*: Australia & Pacific Science Foundation, "Tracing antiquity of banana cultivation in Papua New Guinea," AP Science, http://www.apscience.org.au/pbf_02_3/

*"On the left, a peach as it was"*: Genetic Literacy Project, "How your food would look if not genetically modified over millennia," GLP, 2014, https://geneticliteracyproject.org/2014/06/19/ how-your-food-would-look-if-not-genetically-modified-over-millennia/

*"Peach [4]"* by Rick Harris is licensed with CC BY-SA 2.0.

*"Spanish Cherries"* by leguico is licensed with CC BY-NC-ND 2.0.

*"Sweets, such as jelly beans"*: U.S. Department of Agriculture, "Candies, jellybeans," FoodData Central, 2019, accessed August 30th, 2019, https:// fdc.nal.usda.gov/fdc-app.html#/food-details/167991/nutrients

*"Fruit, such as cherries"*: U.S. Department of Agriculture, "Cherries, sweet, raw," FoodData Central, 2019, accessed August 30th, 2019, https://fdc.nal.usda. gov/fdc-app.html#/food-details/171719/nutrients

*"Even tomatoes have been turned"*: U.S. Department of Agriculture, "Tomato, roma," FoodData Central, 2021, accessed August 30th, 2019, https://fdc. nal.usda.gov/fdc-app.html#/food-details/1750354/nutrients

*"ketchup"*: U.S. Department of Agriculture, "Ketchup, restaurant," FoodData Central, 2019, accessed August 30th, 2019, https://fdc.nal.usda.gov/ fdc-app.html#/food-details/747693/nutrients

*"Sugar has become ever more concentrated"*: Robert Lustig, *Metabolical: The Lure and the Lies of Processed Food, Nutrition, and Modern Medicine* (New York:

Harper Wave, 2021).

*"it's hard for our brain to curb its cravings"*: Kevin Hall et al., "Ultra-processed diets cause excess calorie intake and weight gain: an inpatient randomized controlled trial of ad libitum food intake," *Cell Metabolism* 30, no. 1 (2019): 67-77, https://www.cell.com/action/showPdf?p ii=S1550-4131(19)30248-7

*"it gives us pleasure"*: Robert Lustig, *The Hacking of the American Mind: The Science behind the Corporate Takeover of our Bodies and Brains"* (New York: Penguin, 2017).

## Chapter 5: Underneath our skin: discovering glucose spikes

*"The American Diabetes Association (ADA) states that"*: American Diabetes Association, "Understanding A1C: Diagnosis," Diabetes, accessed August 30th, 2019, https://www.diabetes.org/a1c/diagnosis

*"more likelihood of developing health problems from 4.7 mmol/L"*: Jørgen Bjørnholt et al., "Fasting blood glucose: an underestimated risk factor for cardiovascular death. Results from a 22-year follow-up of healthy nondiabetic men," *Diabetes Care* 22, no. 1 (1999): 45-49, https://care. diabetesjournals.org/content/22/1/45

*"more likelihood of developing health problems from 4.7 mmol/L"*: Chanshin Park et al., "Fasting glucose level and the risk of incident atherosclerotic cardiovascular diseases," *Diabetes Care* 36, no. 7 (2013): 1988-1993, https://care.diabetesjournals.org/content/36/7/1988

*"more likelihood of developing health problems from 4.7 mmol/L"*: Quoc Manh Nguyen et al,. "Fasting plasma glucose levels within the normoglycemic range in childhood as a predictor of prediabetes and type 2 diabetes in adulthood: the Bogalusa Heart Study," *Archives of Pediatrics & Adolescent medicine* 164, no. 2 (2010): 124-128, https://jamanetwork.com/journals/ jamapediatrics/fullarticle/382778

*"Studies in nondiabetics give more precise information"*: Guido Freckmann et al., "Continuous glucose profiles in healthy subjects under everyday life conditions and after different meals," *Journal of Diabetes Science and Technology* 1, no. 5 (2007): 695-703, https://www.ncbi.nlm.nih.gov/pmc/ articles/PMC2769652/

*"it's the* variability *caused by spikes that is most problematic"*: Antonio Ceriello et al., "Oscillating glucose is more deleterious to endothelial function and oxidative stress than mean glucose in normal and type 2 diabetic patients," *Diabetes* 57, no.5 (2008): 1349-1354, https://diabetes.diabetesjournals.org/ content/57/5/1349.short

*"it's the* variability *caused by spikes that is most problematic"*: Louis Monnier et al., "Activation of oxidative stress by acute glucose fluctuations compared with sustained chronic hyperglycemia in patients with type 2 diabetes." *JAMA* 295, no. 14 (2006): 1681-1687. https://jamanetwork.com/journals/ jama/article-abstract/202670

*"it's the* variability *caused by spikes that is most problematic"*: Giada Acciaroli et al., "Diabetes and prediabetes classification using glycemic variability

indices from continuous glucose monitoring data." *Journal of Diabetes Science and Technology* 12, no. 1 (2018): 105-113, https://www.ncbi.nlm. nih.gov/pmc/articles/PMC5761967/

*"The smaller your glycaemic variability"*: Zheng Zhou et al., "Glycemic variability: adverse clinical outcomes and how to improve it?," *Cardiovascular Diabetology* 19, no. 1 (2020): 1-14, https://link.springer. com/article/10.1186/s12933-020-01085-6


## PART II: WHY ARE GLUCOSE SPIKES HARMFUL?

### Chapter 6: Trains, toast and Tetris: the three things that happen in our body when we spike

*"more than 30 trillion cells"*: Ron Sender et al., "Revised estimates for the number of human and bacteria cells in the body," *PLoS Biology* 14, no. 8 (2016): e1002533, https://journals.plos.org/plosbiology/article?id=10.1371/ journal.pbio.1002533

*"the Allostatic Load Model"*: Martin Picard et al., "Mitochondrial allostatic load puts the 'gluc' back in glucocorticoids," *Nature Reviews Endocrinology* 10, no. 5 (2014): 303-310, https://www.uclahealth.org/reversibility-network/ workfiles/resources/publications/picard-endocrinol.pdf

*"free radicals"*: Biplab Giri et al., "Chronic hyperglycemia mediated physiological alteration and metabolic distortion leads to organ dysfunction, infection, cancer progression and other pathophysiological consequences: an update on glucose toxicity," *Biomedicine & Pharmacotherapy*, no. 107 (2018): 306-328, https://www.sciencedirect. com/science/article/pii/S0753332218322406#fig0005

*"Oxidative stress is a driver of"*: Martin Picard et al., "Mitochondrial allostatic load puts the 'gluc' back in glucocorticoids," *Nature Reviews Endocrinology* 10, no. 5 (2014): 303-310, https://www.uclahealth.org/reversibility- network/workfiles/resources/publications/picard-endocrinol.pdf

*"fructose increases oxidative stress even more"*: Robert H. Lustig, "Fructose: it's 'alcohol without the buzz'," *Advances in nutrition* 4, no. 2 (2013): 226-235, https://www.ncbi.nlm.nih.gov/pmc/articles/PMC3649103/

*"Too much fat can also increase oxidative stress"*: Joseph Evans et al., "Are oxidative stress – activated signaling pathways mediators of insulin resistance and β-cell dysfunction?," *Diabetes* 52, no. 1 (2003): 1-8, https:// diabetes.diabetesjournals.org/content/52/1/1.short

*"you are browning, just like a slice of bread"*: Jaime Uribarri et al., "Advanced glycation end products in foods and a practical guide to their reduction in the diet," *Journal of the American Dietetic Association* 100, no. 6 (2010): 911-916, https://www.ncbi.nlm.nih.gov/pmc/articles/PMC3704564/

*"When scientists look at the rib cage cartilage"*: D. G. Dyer et al., "The Maillard reaction in vivo," *Zeitschrift für Ernährungswissenschaft* 30, no. 1 (1991): 29-45, https://www.researchgate.net/ publication/21298410_The_Maillard_reaction_in_vivo.

*"This process is a normal and inevitable part of life"*: Chan-Sik Kim et al., "The

role of glycation in the pathogenesis of aging and its prevention through herbal products and physical exercise," *Journal of Exercise Nutrition & Biochemistry* 21, no. 3 (2017): 55, https://www.ncbi.nlm.nih.gov/pmc/articles/PMC5643203

*"consequences of glycated molecules range from wrinkles"*: Masamitsu Ichihashi et al., "Glycation stress and photo-aging in skin," *Anti-Aging Medicine* 8, no. 3 (2011): 23-29, https://www.jstage.jst.go.jp/article/jaam/8/3/8_3_23/_article/-char/ja/

*"cataracts"*: Ashok Katta et al., "Glycation of lens crystalline protein in the pathogenesis of various forms of cataract," *Biomedical Research* 20, no. 2 (2009): 119-21, https://www.researchgate.net/profile/Ashok-Katta-3/publication/233419577_Glycation_of_lens_crystalline_protein_in_the_pathogenesis_of_various_forms_of_cataract/links/02e7e531342066c955000000/Glycation-of-lens-crystalline-protein-in-the-pathogenesis-of-various-forms-of-cataract.pdf

*"heart disease"*: Georgia Soldatos et al., "Advanced glycation end products and vascular structure and function," *Current Hypertension reports* 8, no. 6 (2006): 472-478, https://pubmed.ncbi.nlm.nih.gov/17087858/

*"Alzheimer's"*: Masayoshi Takeuchi et al., "Involvement of advanced glycation end-products (AGEs) in Alzheimer's disease," *Current Alzheimer Research* 1, no. 1 (2004): 39-46, https://www.ingentaconnect.com/content/ben/car/2004/00000001/00000001/art00006

*"slowing down the browning reaction in your body leads to a longer life"*: Chan-Sik Kim et al., "The role of glycation in the pathogenesis of aging and its prevention through herbal products and physical exercise," *Journal of Exercise Nutrition & Biochemistry* 21, no. 3 (2017): 55, https://www.ncbi.nlm.nih.gov/pmc/articles/PMC5643203

*"Fructose molecules glycate things 10 times"*: Alejandro Gugliucci, "Formation of fructose-mediated advanced glycation end products and their roles in metabolic and inflammatory diseases," *Advances in Nutrition* 8, no. 1 (2017): 54-62, https://www.ncbi.nlm.nih.gov/pmc/articles/PMC5227984/.

*"Fructose molecules glycate things 10 times"*: Alejandro Gugliucci, "Formation of fructose-mediated advanced glycation end products and their roles in metabolic and inflammatory diseases," *Advances in nutrition* 8, no. 1 (2017): 54-62, https://www.ncbi.nlm.nih.gov/pmc/articles/PMC5227984/

*"inflammation-based diseases 'the greatest threat to human health'."*: Roma Pahwa et al., "Chronic inflammation," (2018), https://www.ncbi.nlm.nih.gov/books/NBK493173/

*"three out of five people will die of an inflammation-based disease"*: Roma Pahwa et al., "Chronic inflammation," (2018), https://www.ncbi.nlm.nih.gov/books/NBK493173/

*"Glycogen is actually the cousin of starch"*: "The bonds are also alpha-1,4-glycosidic bond," In: *Biochemistry, 5th ed.* (New York: W. H. Freeman and Co., 1995).

*"liver can hold about 100 grams of glucose"*: David H. Wasserman, "Four grams of glucose," *American Journal of Physiology-Endocrinology and Metabolism*

296, no. 1 (2009): E11-E21, https://www.ncbi.nlm.nih.gov/pmc/articles/PMC2636990/

*"That's half of the 200 grams"*: Jeremy M. Berg, *Biochemistry, 5th edition* (New York: W. H. Freeman, 2002), Section 30.2, https://www.ncbi.nlm.nih.gov/books/NBK22436/#:~:text=The%20brain%20lacks%20fuel%20stores,body%20in%20the%20resting%20state.

*"The muscles of a typical 150-pound adult can hold about 400 grams of glucose"*: David H. Wasserman, "Four grams of glucose," *American Journal of Physiology-Endocrinology and Metabolism* 296, no. 1 (2009): E11-E21, https://www.ncbi.nlm.nih.gov/pmc/articles/PMC2636990/

*"any excess glucose is turned into fat"*: Lubert Stryer, "Fatty acid metabolism," In: *Biochemistry, 4th edition*, (New York: W. H. Freeman and Co., 1995), pp. 603-628.

*"The only thing that fructose can be stored as is fat"*: Samir Softic et al., "Role of dietary fructose and hepatic de novo lipogenesis in fatty liver disease," *Digestive Diseases and Sciences* 61, no. 5 (2016): 1282-1293, https://www.ncbi.nlm.nih.gov/pmc/articles/PMC4838515/

*"it accumulates in the liver"*: Bettina Geidl-Flueck et al., "Fructose-and sucrose-but not glucose-sweetened beverages promote hepatic de novo lipogenesis: A randomized controlled trial," *Journal of Hepatology* 75, no. 1 (2021): 46-54, https://www.journal-of-hepatology.eu/article/S0168-8278(21)00161-6/fulltext#%20

*"The absence of fructose means that fewer molecules end up as fat"*: João Silva et al., "Determining contributions of exogenous glucose and fructose to de novo fatty acid and glycerol synthesis in liver and adipose tissue." *Metabolic Engineering* 56 (2019): 69-76, https://www.sciencedirect.com/science/article/pii/S109671761930196X#fig5

*"The more you're able to grow the number and size of your fat cells"*: Benjamin Bikman, *Why We Get Sick: The Hidden Epidemic at the Root of Most Chronic Disease and How to Fight It* (New York: BenBella, 2020).

*"when our glycogen reserves begin to diminish, our body draws on the fat in our fat reserves for energy"*: Lubert Stryer, *Biochemistry*, 5th ed. (New York: W. H. Freeman and Co., 1995), 773–74.

*"weight loss is always preceded by insulin decrease"*: Natasha Wiebe et al., "Temporal associations among body mass index, fasting insulin, and systemic inflammation: a systematic review and meta-analysis." *JAMA network open* 4, no. 3 (2021): e211263-e211263, https://jamanetwork.com/journals/jamanetworkopen/fullarticle/2777423

**Chapter 7: From head to toe: how spikes make us sick**

*"short-term symptoms associated with spikes"*: Martin Picard et al., "Mitochondrial allostatic load puts the 'gluc' back in glucocorticoids," *Nature Reviews Endocrinology* 10, no. 5 (2014): 303-310, https://www.uclahealth.org/reversibility-network/workfiles/resources/publications/picard-endocrinol.pdf

**Constant hunger**

*"if you compare two meals"*: Paula Chandler-Laney et al., "Return of hunger following a relatively high carbohydrate breakfast is associated with earlier recorded glucose peak and nadir," *Appetite* 80 (2014): 236-241, https://www.sciencedirect.com/science/article/abs/pii/S0195666314002049

*"constant hunger is a symptom of high insulin levels"*: Benjamin Bikman, *Why We Get Sick: The Hidden Epidemic at the Root of Most Chronic Disease and How to Fight It* (New York: BenBella, 2020).

**Cravings**

*"an experiment that took place"*: Kathleen Page et al., "Circulating glucose levels modulate neural control of desire for high-calorie foods in humans," *The Journal of Clinical Investigation* 121, no. 10 (2011): 4161-4169, https://www.jci.org/articles/view/57873

**Chronic fatigue**

*"people born with mitochondrial defects"*: Tanja Taivassalo et al., "The spectrum of exercise tolerance in mitochondrial myopathies: a study of 40 patients," *Brain* 126, no. 2 (2003): 413-423, https://pubmed.ncbi.nlm.nih.gov/12538407/

*"Difficult events, whether physical or mental"*: Martin Picard et al., "Mitochondrial allostatic load puts the 'gluc' back in glucocorticoids," *Nature Reviews Endocrinology*, no. 10.5 (2014): 303-310, https://www.uclahealth.org/reversibility-network/workfiles/resources/publications/picard-endocrinol.pdf

*"impairing the long-term ability of our mitochondria"*: Martin Picard et al., "Mitochondrial allostatic load puts the 'gluc' back in glucocorticoids," *Nature Reviews Endocrinology*, no. 10.5 (2014): 303-310, https://www.uclahealth.org/reversibility-network/workfiles/resources/publications/picard-endocrinol.pdf

*"Diets that cause glucose roller coasters"*: Kara L. Breymeyer et al., "Subjective mood and energy levels of healthy weight and overweight/obese healthy adults on high-and low-glycemic load experimental diets," *Appetite* 107 (2016): 253-259, https://pubmed.ncbi.nlm.nih.gov/27507131/

**Poor sleep**

*"Going to sleep with"*: James Gangwisch et al., "High glycemic index and glycemic load diets as risk factors for insomnia: analyses from the Women's Health Initiative," *The American Journal of Clinical Nutrition* 111, no. 2 (2020): 429-439, https://pubmed.ncbi.nlm.nih.gov/31828298/

*"sleep apnoea"*: R. N. Aurora et al., "Obstructive Sleep Apnea and Postprandial Glucose Differences in Type 2 Diabetes Mellitus," In *A97. SRN: New insights into the cardiometabolic consequences of insufficient sleep*, pp. A2525-A2525. American Thoracic Society, 2020, https://www.atsjournals.org/doi/abs/10.1164/ajrccm-conference.2020.201.1_MeetingAbstracts.A2525

**Colds and coronavirus complications**

*"immune system is temporarily faulty"*: Nagham Jafar et al., "The effect of short-term hyperglycemia on the innate immune system," *The American Journal of the Medical Sciences* 351, no. 2 (2016): 201-211, https://www.amjmedsci. org/article/S0002-9629(15)00027-0/fulltext

*"goodbye to five-star immune responses"*: Janan Kiselar et al., "Modification of β-Defensin-2 by dicarbonyls methylglyoxal and glyoxal inhibits antibacterial and chemotactic function in vitro," *PLoS One* 10, no. 8 (2015): e0130533, https://journals.plos.org/plosone/article?id=10.1371/ journal.pone.0130533

*"main determinants of recovery from a coronavirus infection"*: Jiaoyue Zhang et al., "Impaired fasting glucose and diabetes are related to higher risks of complications and mortality among patients with coronavirus disease 2019," *Frontiers in Endocrinology* 11 (2020): 525, https://www.frontiersin. org/articles/10.3389/fendo.2020.00525/full?report=reader

*"more easily infected"*: Emmanuelle Logette et al., "A Machine-Generated View of the Role of Blood Glucose Levels in the Severity of COVID-19," *Frontiers in Public Health* (2021): 1068, https://www.frontiersin.org/ articles/10.3389/fpubh.2021.695139/full?fbclid=IwAR0RS9OVCuL9q-fbW4gF7McCYfgRgNDQIVI4JjZE-59Sm1E7l1MFZ0ZGyoI

*"twice as likely to die from the virus"*: Francisco Carrasco-Sánchez et al., "Admission hyperglycaemia as a predictor of mortality in patients hospitalized with COVID-19 regardless of diabetes status: data from the Spanish SEMI-COVID-19 Registry," *Annals of Medicine* 53, no. 1 (2021): 103-116, https://www.tandfonline.com/doi/full/10.1080/07853890.2020. 1836566

**Gestational diabetes is harder to manage**

*"That's because insulin"*: Ursula Hiden et al., "Insulin and the IGF system in the human placenta of normal and diabetic pregnancies," *Journal of Anatomy* 215, no. 1 (2009): 60-68, https://onlinelibrary.wiley.com/doi/ full/10.1111/j.1469-7580.2008.01035.x

*"mom's breast tissue"*: Chiara Berlato et al., "Selective response to insulin versus insulin-like growth factor-I and -II and up-regulation of insulin receptor splice variant B in the differentiated mouse mammary epithelium," *Endocrinology* 150, no. 6 (2009): 2924-2933, https://academic.oup.com/ endo/article/150/6/2924/2456369?login=true

*"mothers can reduce their likelihood of needing medication"*: Carol Major et al., "The effects of carbohydrate restriction in patients with diet-controlled gestational diabetes," *Obstetrics & Gynecology* 91, no. 4 (1998): 600-604, https://www.sciencedirect.com/science/article/abs/pii/ S0029784498000039

*"reduce the likelihood of a cesarean"*: Robert Moses et al., "Effect of a low-glycemic-index diet during pregnancy on obstetric outcomes," *The American Journal of Clinical Nutrition* 84, no. 4 (2006): 807-812, https:// academic.oup.com/ajcn/article/84/4/807/4633214

*"limit their own weight gain"*: James F. Clapp III et al., "Maternal carbohydrate intake and pregnancy outcome," *Proceedings of the Nutrition Society* 61, no. 1 (2002): 45-50, https://www.cambridge.org/core/journals/proceedings-of-the-nutrition-society/article/maternal-carbohydrate-intake-and-pregnancy-outcome/28F8E1C5E1460E67F2F1CE0C1D06EE81

## Hot flashes and night sweats
*"Research shows that the symptoms of menopause"*: Rebecca Thurston et al., "Vasomotor symptoms and insulin resistance in the study of women's health across the nation," *Journal of Clinical Endocrinology & Metabolism* 97, no. 10 (2012): 3487-3494, https://pubmed.ncbi.nlm.nih.gov/22851488/
*"a 2020 study from Columbia University"*: James E Gangwisch et al., "High glycemic index and glycemic load diets as risk factors for insomnia: analyses from the Women's Health Initiative," *American Journal of Clinical Nutrition* 111, no. 2 (2020): 429-439, https://pubmed.ncbi.nlm.nih.gov/31828298/

## Migraine
*"data proves that women with insulin resistance"*: A. Fava et al., "Chronic migraine in women is associated with insulin resistance: a cross-sectional study," *European Journal of Neurology* 21, no. 2 (2014): 267-272, https://onlinelibrary.wiley.com/doi/abs/10.1111/ene.12289
*"When sufferers' insulin levels are lowered"*: Cinzia Cavestro et al., "Alpha-lipoic acid shows promise to improve migraine in patients with insulin resistance: a 6-month exploratory study," *Journal of Medicinal Food* 21, no. 3 (2018): 269-273, https://www.liebertpub.com/doi/abs/10.1089/jmf.2017.0068

## Memory and cognitive function
*"big glucose spikes can impair"*: Rachel Ginieis et al., "The 'sweet' effect: comparative assessments of dietary sugars on cognitive performance," *Physiology & Behavior* 184 (2018): 242-247, https://pubmed.ncbi.nlm.nih.gov/29225094/
*"worst first thing in the morning"*: Rachel Ginieis et al., "The "sweet" effect: comparative assessments of dietary sugars on cognitive performance," *Physiology & behavior* 184 (2018): 242-247, https://pubmed.ncbi.nlm.nih.gov/29225094/

## Acne and other skin conditions
*"starchy and sugary foods"*: Hyuck Hoon Kwon et al., "Clinical and histological effect of a low glycaemic load diet in treatment of acne vulgaris in Korean patients: a randomized, controlled trial," *Acta dermato-venereologica* 92, no. 3 (2012): 241-246, https://pubmed.ncbi.nlm.nih.gov/22678562/
*"acne clears up"*: Robyn N. Smith et al., "A low-glycemic-load diet improves symptoms in acne vulgaris patients: a randomized controlled trial," *American Journal of Clinical Nutrition* 86, no. 1 (2007): 107-115, https://

pubmed.ncbi.nlm.nih.gov/17616769/

**Ageing and arthritis**

*"The more often we spike, the faster we age"*: George Suji et al., "Glucose, glycation and aging," *Biogerontology* 5, no. 6 (2004): 365-373, https://link. springer.com/article/10.1007/s10522-004-3189-0

*"Glycation, free radicals, and the subsequent inflammation"*: Roma Pahwa et al., "Chronic inflammation," (2018), https://www.ncbi.nlm.nih.gov/books/ NBK493173/

*"rheumatoid arthritis"* : Roma Pahwa et al., "Chronic inflammation," (2018), https://www.ncbi.nlm.nih.gov/books/NBK493173/

*"degradation of cartilage"*: Robert A. Greenwald et al., "Inhibition of collagen gelation by action of the superoxide radical," *Arthritis & Rheumatism: Official Journal of the American College of Rheumatology* 22, no. 3 (1979): 251-259, https://pubmed.ncbi.nlm.nih.gov/217393/

*"osteoarthritis"*: Biplab Giri et al., "Chronic hyperglycemia mediated physiological alteration and metabolic distortion leads to organ dysfunction, infection, cancer progression and other pathophysiological consequences: an update on glucose toxicity," *Biomedicine & Pharmacotherapy*, no. 107 (2018): 306-328, https://www.sciencedirect. com/science/article/abs/pii/S0753332218322406

*"our bones waste away"*: John Tower, "Programmed cell death in aging," *Ageing research reviews* 23 (2015): 90-100, https://www.ncbi.nlm.nih.gov/pmc/ articles/PMC4480161/

**Alzheimer's and dementia**

*"The neurons in our brain feel oxidative stress"*: Charles Watt et al., "Glycemic variability and CNS inflammation: Reviewing the connection," *Nutrients* 12, no. 12 (2020): 3906, https://pubmed.ncbi.nlm.nih.gov/33371247/

*"key factor in almost all chronic degenerative diseases"*: Roma Pahwa et al., "Chronic inflammation," (2018), https://www.ncbi.nlm.nih.gov/books/ NBK493173/

*"Alzheimer's is sometimes called 'type 3 diabetes'"*: Suzanne M. De la Monte et al., "Alzheimer's disease is type 3 diabetes – evidence reviewed," *Journal of Diabetes Science and Technology* 2, no. 6 (2008): 1101-1113, https:// journals.sagepub.com/doi/abs/10.1177/193229680800200619

*"type 2 diabetics are four times as likely to develop Alzheimer's"*: Robert H. Lustig, *Metabolical: The Lure and the Lies of Processed Food, Nutrition, and Modern Medicine* (New York: Harper Wave, 2021).

*"The signs are visible early, too"*: Jiyin Zhou et al., "Diabetic cognitive dysfunction: from bench to clinic," *Current Medicinal Chemistry* 27, no. 19 (2020): 3151-3167, https://pubmed.ncbi.nlm.nih.gov/30727866/

*"The signs are visible early, too"*: Auriel A. Willette et al., "Association of insulin resistance with cerebral glucose uptake in late middle–aged adults at risk for Alzheimer disease," *JAMA neurology* 72, no. 9 (2015): 1013-1020, https://pubmed.ncbi.nlm.nih.gov/26214150/

*"The signs are visible early, too"*: Christine M. Burns et al., "Higher serum glucose levels are associated with cerebral hypometabolism in Alzheimer regions," *Neurology* 80, no. 17 (2013): 1557-1564, https://www.ncbi.nlm. nih.gov/pmc/articles/PMC3662330/

*"studies show short-term"*: Mark A. Reger et al., "Effects of β-hydroxybutyrate on cognition in memory-impaired adults," *Neurobiology of Aging* 25, no. 3 (2004): 311-314, https://www.sciencedirect.com/science/article/abs/pii/ S0197458003000873

*"and long-term improvements"*: Dale E Bredesen, et al., "Reversal of cognitive decline: a novel therapeutic program," *Aging (Albany NY)* 6, no. 9 (2014): 707, https://www.ncbi.nlm.nih.gov/pmc/articles/PMC4221920/

*"A therapeutic program out of UCLA"*: Dale E. Bredesen, et al., "Reversal of cognitive decline: a novel therapeutic program," *Aging (Albany NY)* 6, no. 9 (2014): 707, https://www.ncbi.nlm.nih.gov/pmc/articles/PMC4221920/

**Cancer risk**

*"Children born today"*: Amar S. Ahmad et al., "Trends in the lifetime risk of developing cancer in Great Britain: comparison of risk for those born from 1930 to 1960." *British Journal of Cancer* 112, no. 5 (2015): 943-947, https:// www.nature.com/articles/bjc2014606

*"poor diet, together with smoking, is the main driver"*: Robert H. Lustig, *Metabolical: The Lure and the Lies of Processed Food, Nutrition, and Modern Medicine* (New York: Harper Wave, 2021).

*"cancer may begin with"*: Florian R. Greten et al., "Inflammation and cancer: triggers, mechanisms, and consequences," *Immunity* 51, no. 1 (2019): 27-41, https://www.sciencedirect.com/science/article/pii/ S107476131930295X

*"when there is more insulin present, cancer spreads even faster"*: Rachel J. Perry et al., "Mechanistic links between obesity, insulin, and cancer," *Trends in Cancer* 6, no. 2 (2020): 75-78, https://www.sciencedirect.com/science/ article/abs/pii/S2405803319302614

*"Glucose is the key to many of these processes"*: Tetsuro Tsujimoto et al., "Association between hyperinsulinemia and increased risk of cancer death in nonobese and obese people: A population-based observational study," *International Journal of Cancer* 141, no. 1 (2017): 102-111, https:// onlinelibrary.wiley.com/doi/full/10.1002/ijc.30729

**Depressive episodes**

*"When people eat a diet that"*: Kara L. Breymeyer et al., "Subjective mood and energy levels of healthy weight and overweight/obese healthy adults on high-and low-glycemic load experimental diets," *Appetite* 107 (2016): 253-259, https://pubmed.ncbi.nlm.nih.gov/27507131/

*"worsening moods, more depressive symptoms"*: Rachel A. Cheatham et al., "Long-term effects of provided low and high glycemic load low energy diets on mood and cognition," *Physiology & Behavior* 98, no. 3 (2009): 374-379, https://pubmed.ncbi.nlm.nih.gov/19576915/

"*worsening moods, more depressive symptoms*": Sue Penckofer et al., "Does glycemic variability impact mood and quality of life?," *Diabetes Technology & Therapeutics* 14, no. 4 (2012): 303-310, https://www.ncbi.nlm.nih.gov/pmc/articles/PMC3317401/

"*symptoms get worse as the spikes get more extreme*": James E. Gangwisch et al., "High glycemic index diet as a risk factor for depression: analyses from the Women's Health Initiative," *American Journal of Clinical Nutrition* 102, no. 2 (2015): 454-463, https://www.ncbi.nlm.nih.gov/pmc/articles/PMC4515860/

**Gut issues**

"*high glucose levels, for example, could increase leaky gut syndrome*": Fernando F. Anhê et al., "Glucose alters the symbiotic relationships between gut microbiota and host physiology," *American Journal of Physiology-Endocrinology and Metabolism* 318, no. 2 (2020): E111-E116, https://pubmed.ncbi.nlm.nih.gov/31794261/

"*food allergies and other autoimmune diseases*": Robert H. Lustig, *Metabolical: The Lure and the Lies of Processed Food, Nutrition, and Modern Medicine* (New York: Harper Wave, 2021).

"*can get rid of their heartburn or acid reflux*": William S. Yancy et al., "Improvements of gastroesophageal reflux disease after initiation of a low-carbohydrate diet: Five brief case reports," *Alternative Therapies in Health and Medicine* 7, no. 6 (2001): 120, https://search.proquest.com/openview/1c418d7f0548f58a5c647b1204d3f6a7/1?pq-origsite=gscholar&cbl=32528

"*gut health is linked to mental health*": Jessica M. Yano et al., "Indigenous bacteria from the gut microbiota regulate host serotonin biosynthesis," *Cell* 161, no. 2 (2015): 264-276, https://www.ncbi.nlm.nih.gov/pmc/articles/PMC4393509/

"*gut health is linked to mental health*": Roberto Mazzoli et al., "The neuro-endocrinological role of microbial glutamate and GABA signaling," *Frontiers in Microbiology* 7 (2016): 1934, https://www.ncbi.nlm.nih.gov/pmc/articles/PMC5127831/

"*The gut and the brain are connected*": Emeran A. Mayer, "Gut feelings: the emerging biology of gut-brain communication," *Nature Reviews Neuroscience* 12, no. 8 (2011): 453-466, https://www.ncbi.nlm.nih.gov/pmc/articles/PMC3845678/

"*Information is sent back and forth between them*": Sigrid Breit et al., "Vagus nerve as modulator of the brain–gut axis in psychiatric and inflammatory disorders," *Frontiers in Psychiatry* 9 (2018): 44, https://www.ncbi.nlm.nih.gov/pubmed/29593576

"*Information is sent back and forth between them*": Bruno Bonaz et al., "The vagus nerve at the interface of the microbiota-gut-brain axis," *Frontiers in Neuroscience* 12 (2018): 49, https://www.ncbi.nlm.nih.gov/pubmed/29467611

## Heart disease

*"half of the people who have a heart attack have normal levels of cholesterol"*:
Michael D. Miedema et al., "Statin eligibility and outpatient care prior
to ST-segment elevation myocardial infarction," *Journal of the American
Heart Association* 6, no. 4 (2017): e005333, https://www.ahajournals.org/
doi/10.1161/JAHA.116.005333

*"our liver starts producing LDL pattern B"*: Benjamin Bikman, *Why We Get Sick:
The Hidden Epidemic at the Root of Most Chronic Disease and How to Fight
It* (New York: BenBella, 2020).

*"if and when that cholesterol is oxidised"*: Benjamin Bikman, *Why We Get Sick:
The Hidden Epidemic at the Root of Most Chronic Disease and How to Fight
It* (New York: BenBella, 2020).

*"each additional glucose spike increases our risk of dying of a heart attack"*: Koichi
Node et al., "Postprandial hyperglycemia as an etiological factor in
vascular failure," *Cardiovascular Diabetology* 8, no. 1 (2009): 1-10, https://
pubmed.ncbi.nlm.nih.gov/19402896/

*"each additional glucose spike increases our risk of dying of a heart attack"*: Antonio
Ceriello et al., "Oscillating glucose is more deleterious to endothelial
function and oxidative stress than mean glucose in normal and type 2
diabetic patients," *Diabetes* 57, no. 5 (2008): 1349-1354, https://pubmed.
ncbi.nlm.nih.gov/18299315/

*"each additional glucose spike increases our risk of dying of a heart attack"*:
Michelle Flynn et al., "Transient intermittent hyperglycemia accelerates
atherosclerosis by promoting myelopoiesis," *Circulation Research* 127,
no. 7 (2020): 877-892, https://www.ahajournals.org/doi/full/10.1161/
CIRCRESAHA.120.316653

*"each additional glucose spike increases our risk of dying of a heart attack"*: E.
Succurro et al., "Elevated one-hour post-load plasma glucose levels
identifies subjects with normal glucose tolerance but early carotid
atherosclerosis," *Atherosclerosis* 207, no. 1 (2009): 245-249, https://www.
sciencedirect.com/science/article/abs/pii/S0021915009002718

*"statins lower LDL pattern A"*: Benjamin Bikman, *Why We Get Sick: The Hidden
Epidemic at the Root of Most Chronic Disease and How to Fight It* (New
York: BenBella, 2020).

*"statins don't decrease the risk of a first heart attack"*: Robert H. Lustig,
*Metabolical: The Lure and the Lies of Processed Food, Nutrition, and Modern
Medicine* (New York: Harper Wave, 2021).

*"a ratio that is surprisingly accurate in predicting LDL size"*: Benjamin Bikman,
*Why We Get Sick: The Hidden Epidemic at the Root of Most Chronic Disease
and How to Fight It* (New York: BenBella, 2020).

*"measuring C-reactive protein"*: Paul M Ridker et al., "Comparison of C-reactive
protein and low-density lipoprotein cholesterol levels in the prediction
of first cardiovascular events," *New England Journal of Medicine* 347,
no. 20 (2002): 1557-1565, https://www.nejm.org/doi/full/10.1056/
NEJMoa021993

**Infertility and polycystic ovarian syndrome (PCOS)**

*"women and men with high insulin levels are more likely to be infertile"*: Tetsurou Sakumoto et al., "Insulin resistance/hyperinsulinemia and reproductive disorders in infertile women," *Reproductive Medicine and Biology* 9, no. 4 (2010): 185-190, https://www.ncbi.nlm.nih.gov/pmc/articles/PMC5904600/

*"women and men with high insulin levels are more likely to be infertile"*: LaTasha B. Craig et al., "Increased prevalence of insulin resistance in women with a history of recurrent pregnancy loss," *Fertility and Sterility* 78, no. 3 (2002): 487-490, https://www.sciencedirect.com/science/article/abs/pii/S0015028202032478

*"women and men with high insulin levels are more likely to be infertile"*: Nelly Pitteloud et al., "Increasing insulin resistance is associated with a decrease in Leydig cell testosterone secretion in men," *Journal of Clinical Endocrinology & Metabolism* 90, no. 5 (2005): 2636-2641, https://academic.oup.com/jcem/article/90/5/2636/2836773

*"The more glucose spikes in our diet"*: Jorge E. Chavarro et al., "A prospective study of dietary carbohydrate quantity and quality in relation to risk of ovulatory infertility," *European Journal of Clinical Nutrition* 63, no. 1 (2009): 78-86, https://www.ncbi.nlm.nih.gov/pmc/articles/PMC3066074/

*"polycystic ovarian syndrome (PCOS)"*: Centers for Disease Control and Prevention, "PCOS (Polycystic Ovary Syndrome) and Diabetes," CDC, accessed August 30th, 2021, https://www.cdc.gov/diabetes/basics/pcos.html

*"insulin tells the ovaries to produce more testosterone"*: John E. Nestler et al., "Insulin stimulates testosterone biosynthesis by human thecal cells from women with polycystic ovary syndrome by activating its own receptor and using inositolglycan mediators as the signal transduction system," *Journal of Clinical Endocrinology & Metabolism* 83, no. 6 (1998): 2001-2005, https://academic.oup.com/jcem/article/83/6/2001/2865383?login=true

*"On top of that, with too much insulin"*: Benjamin Bikman, *Why We Get Sick: The Hidden Epidemic at the Root of Most Chronic Disease and How to Fight It* (New York: BenBella, 2020).

*"women suffering from PCOS display masculine traits"*: Centers for Disease Control and Prevention, "PCOS (Polycystic Ovarian Syndrome) and Diabetes," CDC, accessed August 30th, 2021, https/www.cdc.gov/diabetes/basics/pcos.html

*"In a study performed at Duke University"*: John C. Mavropoulos et al., "The effects of a low-carbohydrate, ketogenic diet on the polycystic ovary syndrome: a pilot study," *Nutrition & Metabolism* 2, no. 1 (2005): 1-5, https://www.ncbi.nlm.nih.gov/pmc/articles/PMC1334192/

*"For men, dysregulated glucose is also linked to infertility"*: Zeeeshan Anwar et al., "Erectile dysfunction: An underestimated presentation in patients with diabetes mellitus," *Indian Journal of Psychological Medicine* 39, no. 5 (2017): 600-604, https://www.ncbi.nlm.nih.gov/pmc/articles/PMC5688886/

*"erectile dysfunction in men under 40 years"*: Fengjuan Yao et al., "Erectile dysfunction may be the first clinical sign of insulin resistance and endothelial dysfunction in young men," *Clinical Research in Cardiology* 102, no. 9 (2013): 645-651, https://link.springer.com/article/10.1007/s00392-013-0577-y

**Insulin resistance and type 2 diabetes**

*"Type 2 diabetes is a global epidemic"*: Sudesna Chatterjee et al., "Type 2 diabetes," *Lancet* 389, no. 10085 (2017): 2239-2251, https://www.sciencedirect.com/science/article/abs/pii/S0140673617300582

*"more inflammation, a process set off by glucose spikes, makes it worse"*: Marc Y. Donath et al., "Type 2 diabetes as an inflammatory disease," *Nature Reviews Immunology* 11, no. 2 (2011): 98-107, https://pubmed.ncbi.nlm.nih.gov/21233852/

*"most effective way to reverse type 2 diabetes is to flatten our glucose curves."*: Joshua Z. Goldenberg et al., "Efficacy and safety of low and very low carbohydrate diets for type 2 diabetes remission: systematic review and meta-analysis of published and unpublished randomized trial data," *British Medical Journal* 372 (2021), https://www.bmj.com/content/372/bmj.m4743

*"In one study, type 2 diabetics"*: William S. Yancy et al., "A low-carbohydrate, ketogenic diet to treat type 2 diabetes," *Nutrition & Metabolism* 2, no. 1 (2005): 1-7, https://link.springer.com/article/10.1186/1743-7075-2-34

*"American Diabetes Association (the ADA) started endorsing glucose-flattening diets"*: Alison B. Evert et al., "Nutrition therapy for adults with diabetes or prediabetes: a consensus report," *Diabetes Care* 42, no. 5 (2019): 731-754, https://care.diabetesjournals.org/content/diacare/early/2019/04/10/dci19-0014.full.pdf

**Nonalcoholic fatty liver disease**

*"excess fructose could cause liver disease"*: Robert H. Lustig, "Fructose: it's 'alcohol without the buzz'," *Advances in Nutrition* 4, no. 2 (2013): 226-235, https://www.ncbi.nlm.nih.gov/pmc/articles/PMC3649103/

*"one out of every four adults has NAFLD"*: Zobair M. Younossi et al., "Global epidemiology of nonalcoholic fatty liver disease – meta-analytic assessment of prevalence, incidence, and outcomes," *Hepatology* 64, no. 1 (2016): 73-84, https://aasldpubs.onlinelibrary.wiley.com/doi/full/10.1002/hep.28431

*"In people who are overweight, it's even more common"*: Ruth C.R. Meex et al., "Hepatokines: linking nonalcoholic fatty liver disease and insulin resistance." *Nature Reviews Endocrinology* 13, no. 9 (2017): 509-520, https://www.nature.com/articles/nrendo.2017.56

**Wrinkles and cataracts**

*"Broken collagen leads to sagging skin and wrinkles"*: F. William Danby, "Nutrition and aging skin: sugar and glycation," *Clinics in Dermatology* 28, no. 4 (2010): 409-411, https://www.sciencedirect.com/science/article/

· abs/pii/S0738081X10000428

*"The more glycation, the more"*: Paraskevi Gkogkolou et al., "Advanced glycation end products: key players in skin aging?," *Dermato-endocrinology* 4, no. 3 (2012): 259-270, https://www.ncbi.nlm.nih.gov/pmc/articles/PMC3583887/

*"we develop cataracts"*: Ashok V. Katta et al., "Glycation of lens crystalline protein in the pathogenesis of various forms of cataract," *Biomedical Research* 20, no. 2 (2009): 119-21, https://www.researchgate.net/profile/Ashok-Katta-3/publication/233419577_Glycation_of_lens_crystalline_protein_in_the_pathogenesis_of_various_forms_of_cataract/links/02e7e531342066c955000000/Glycation-of-lens-crystalline-protein-in-the-pathogenesis-of-various-forms-of-cataract.pdf

*"Odds are, you are among the 88 per cent of adults who have dysregulated glucose levels"*: Joana Araújo et al., "Prevalence of optimal metabolic health in American adults: National Health and Nutrition Examination Survey 2009–2016," *Metabolic Syndrome and Related Disorders* 17, no. 1 (2019): 46-52, https://www.liebertpub.com/doi/10.1089/met.2018.0105

## PART III: HOW CAN I FLATTEN MY GLUCOSE CURVES?

### Hack 1: Eat foods in the right order

*"if you eat the items of a meal containing starch, fiber, sugar, protein, and fat in a specific order"*: Alpana P. Shukla et al., "Food order has a significant impact on postprandial glucose and insulin levels." *Diabetes Care* 38, no. 7 (2015): e98-e99, https://care.diabetesjournals.org/content/38/7/e98

*"This is true for anyone, with or without diabetes"*: Kimiko Nishino et al., "Consuming carbohydrates after meat or vegetables lowers postprandial excursions of glucose and insulin in nondiabetic subjects," *Journal of Nutritional Science and Vitaminology* 64, no. 5 (2018): 316-320, https://www.researchgate.net/publication/328640463_Consuming_Carbohydrates_after_Meat_or_Vegetables_Lowers_Postprandial_Excursions_of_Glucose_and_Insulin_in_Nondiabetic_Subjects

*"the effect of this sequencing is comparable to the effects of diabetes medications"*: Alpana P. Shukla et al., "Food order has a significant impact on postprandial glucose and insulin levels," *Diabetes Care* 38, no. 7 (2015): e98-e99, https://care.diabetesjournals.org/content/38/7/e98

*"A startling study from 2016 proved the finding even more definitively"*: Domenico Tricò et al., "Manipulating the sequence of food ingestion improves glycemic control in type 2 diabetic patients under free-living conditions," *Nutrition & Diabetes* 6, no. 8 (2016): e226-e226, https://www.nature.com/articles/nutd201633/

*"about three calories' worth of food trickle through"*: Diana Gentilcore et al., "Effects of fat on gastric emptying of and the glycemic, insulin, and incretin responses to a carbohydrate meal in type 2 diabetes," *Journal of Clinical Endocrinology & Metabolism* 91, no. 6 (2006): 2062-2067, https://

academic.oup.com/jcem/article/91/6/2062/2843371?login=true

*"Fibre has three superpowers"*: J. R. Perry et al., "A review of physiological effects of soluble and insoluble dietary fibers," *Journal of Nutrition and Food Sciences* 6, no. 2 (2016): 476, https://www.longdom.org/open-access/a-review-of-physiological-effects-of-soluble-and-insoluble-dietary-fibers-2155-9600-1000476.pdf

*"Foods containing fat also slow down gastric emptying"*: Diana Gentilcore et al., "Effects of fat on gastric emptying of and the glycemic, insulin, and incretin responses to a carbohydrate meal in type 2 diabetes," *The Journal of Clinical Endocrinology & Metabolism* 91, no. 6 (2006): 2062-2067, https://academic.oup.com/jcem/article/91/6/2062/2843371?login=true

*"when we eat foods in the right order, our pancreas produces less insulin"*: Alpana P. Shukla et al., "Food order has a significant impact on postprandial glucose and insulin levels," *Diabetes care* 38, no. 7 (2015): e98-e99, https://care.diabetesjournals.org/content/38/7/e98

*"when we eat foods in the right order, our pancreas produces less insulin"*: Kimiko Nishino et al., "Consuming carbohydrates after meat or vegetables lowers postprandial excursions of glucose and insulin in nondiabetic subjects," *Journal of Nutritional Science and Vitaminology* 64, no. 5 (2018): 316-320, https://www.researchgate.net/publication/328640463_Consuming_Carbohydrates_after_Meat_or_Vegetables_Lowers_Postprandial_Excursions_of_Glucose_and_Insulin_in_Nondiabetic_Subjects

*"ghrelin stays suppressed for much longer"*: Alpana P. Shukla et al., "Effect of food order on ghrelin suppression," *Diabetes Care* 41, no. 5 (2018): e76-e77, https://care.diabetesjournals.org/content/41/5/e76

*"Research also shows that in postmenopausal women"*: James E. Gangwisch et al., "High glycemic index and glycemic load diets as risk factors for insomnia: analyses from the Women's Health Initiative," *American Journal of Clinical Nutrition* 111, no. 2 (2020): 429-439, https://pubmed.ncbi.nlm.nih.gov/31828298/

*"float on top of the contents of the stomach and eventually putrefy"*: David Gentilcore, *Food and Health in Early Modern Europe: Diet, Medicine and Society* (New York: Bloomsbury Publishing, 2015), *1450-1800.*

*"Second, our stomach is an acidic environment"*: R. H. Hunt et al., "The stomach in health and disease," *Gut* 64, no. 10 (2015): 1650-1668, https://www.ncbi.nlm.nih.gov/pmc/articles/PMC4835810/

*"Nothing can rot in the stomach"*: R. H. Hunt et al., "The stomach in health and disease," *Gut* 64, no. 10 (2015): 1650-1668, https://www.ncbi.nlm.nih.gov/pmc/articles/PMC4835810/

*"in Roman times, a meal generally started with"*: Patrick Faas, *Around the Roman Table: food and feasting in ancient Rome* (Chicago: University of Chicago Press, 2005).

**Hack 2: Add a green starter to all your meals**

*"people in the UK don't get nearly enough fibre"*: https://www.nhs.uk/live-well/eat-well/how-to-get-more-fibre-into-your-diet/

*"In the US, only 5 per cent of people"*: Diane Quagliani et al., "Closing America's fiber gap: communication strategies from a food and fiber summit," *American Journal of Lifestyle Medicine* 11, no. 1 (2017): 80-85, https://www.ncbi.nlm.nih.gov/pmc/articles/PMC6124841/

*"nutrient of public health concern"*: United States Dietary Guidelines Advisory Committee, "Dietary guidelines for Americans, 2010," no. 232.

*"This plant-made substance is incredibly important to us"*: Thomas M. Barber et al., "The health benefits of dietary fibre," *Nutrients* 12, no. 10 (2020): 3209, https://www.mdpi.com/2072-6643/12/10/3209/pdf

*"it creates a viscous mesh in our intestine"*: Martin O. Weickert et al., "Metabolic effects of dietary fiber consumption and prevention of diabetes," *Journal of Nutrition* 138, no. 3 (2008): 439-442, https://academic.oup.com/jn/article/138/3/439/4670214

*"the additional fibre reduced the glucose spike of the bread"*: Jannie Yi Fang Yang et al., "The effects of functional fiber on postprandial glycaemia, energy intake, satiety, palatability and gastrointestinal wellbeing: a randomized crossover trial," *Nutrition journal* 13, no. 1 (2014): 1-9, https://nutritionj.biomedcentral.com/articles/10.1186/1475-2891-13-76

*"With a flatter curve, we stay full for longer"*: Paula C. Chandler-Laney et al., "Return of hunger following a relatively high carbohydrate breakfast is associated with earlier recorded glucose peak and nadir," *Appetite* 80 (2014): 236-241, https://www.sciencedirect.com/science/article/abs/pii/S0195666314002049

*"avoid the glucose dip that leads to cravings"*: Patrick Wyatt et al., "Postprandial glycaemic dips predict appetite and energy intake in healthy individuals," *Nature Metabolism* 3, no. 4 (2021): 523-529, https://www.nature.com/articles/s42255-021-00383-x

*"a fibre supplement at the beginning of a meal can help"*: Lorenzo Nesti et al., "Impact of nutrient type and sequence on glucose tolerance: Physiological insights and therapeutic implications," *Frontiers in Endocrinology* 10 (2019): 144, https://www.frontiersin.org/articles/10.3389/fendo.2019.00144/full#B58

*"diabetes isn't just about genes"*: Michael Multhaup at al., *The science behind 23andMe's Type 2 Diabetes report*, 2019, accessed August 30th, 2021, https://permalinks.23andme.com/pdf/23_19-Type2Diabetes_March2019.pdf

*"our lifestyle is still the main reason we do – or don't"*: Michael E.J. Lean et al., "Primary care-led weight management for remission of type 2 diabetes (DiRECT): an open-label, cluster-randomised trial," *Lancet* 391, no. 10120 (2018): 541-551, https://pubmed.ncbi.nlm.nih.gov/29221645/

**Hack 3: Stop counting calories**

*"in 2015, a research team out of UC San Francisco proved that"*: Robert H Lustig et al., "Isocaloric fructose restriction and metabolic improvement in children with obesity and metabolic syndrome," *Obesity* 24, no. 2 (2016): 453-460, https://onlinelibrary.wiley.com/doi/full/10.1002/oby.21371

*"people who focus on flattening their glucose curves can eat more calories"*: Laura
R. Saslow et al., "Twelve-month outcomes of a randomized trial of a
moderate-carbohydrate versus very low-carbohydrate diet in overweight
adults with type 2 diabetes mellitus or prediabetes," *Nutrition & Diabetes*
7, no. 12 (2017): 1-6, https://www.nature.com/articles/s41387-017-0006-9

*"a 2017 study from the University of Michigan"*: Laura R. Saslow et al., "Twelve-
month outcomes of a randomized trial of a moderate-carbohydrate versus
very low-carbohydrate diet in overweight adults with type 2 diabetes
mellitus or prediabetes," *Nutrition & Diabetes* 7, no. 12 (2017): 1-6,
https://www.nature.com/articles/s41387-017-0006-9

*"insulin reduction is primordial and always precedes weight loss"*: Natasha Wiebe
et al., "Temporal associations among body mass index, fasting insulin,
and systemic inflammation: a systematic review and meta-analysis." *JAMA
network open* 4, no. 3 (2021): e211263-e211263, https://jamanetwork.com/
journals/jamanetworkopen/fullarticle/2777423

*"we can completely ignore calories and still lose weight"*: Tian Hu et al.,
"Adherence to low-carbohydrate and low-fat diets in relation to weight loss
and cardiovascular risk factors," *Obesity Science & Practice* 2, no. 1 (2016):
24-31, https://onlinelibrary.wiley.com/doi/full/10.1002/osp4.23

*"Reactive hypoglycaemia is a common condition"*: Hanne Mumm et al.,
"Prevalence and possible mechanisms of reactive hypoglycemia in
polycystic ovary syndrome," *Human Reproduction* 31, no. 5 (2016): 1105-
1112, https://pubmed.ncbi.nlm.nih.gov/27008892/

*"their glucose level can get so low that it causes a coma"*: Gita Shafiee et al., "The
importance of hypoglycemia in diabetic patients," *Journal of Diabetes
& Metabolic Disorders* 11, no. 1 (2012): 1-7, https://link.springer.com/
article/10.1186/2251-6581-11-17

*"the greater the dip, the more hungry we become"*: Patrick Wyatt et al.,
"Postprandial glycaemic dips predict appetite and energy intake in healthy
individuals," *Nature Metabolism* 3, no. 4 (2021): 523-529, https://www.
nature.com/articles/s42255-021-00383-x

**Hack 4: Flatten your breakfast curve**

*"Twenty participants were recruited, both men and women"*: Heather Hall et al.,
"Glucotypes reveal new patterns of glucose dysregulation," *PLoS Biology*
16, no. 7 (2018): e2005143, https://pubmed.ncbi.nlm.nih.gov/30040822/

*"2.7 billion boxes of cereal are sold every year"*: Statista report based on the U.S.
Census data and Simmons National Consumer Survey (NHCS).

*"contains three times as much sugar as the cereal"*: Nutritionix Grocery
Database, "Honey Nut Cheerios, Cereal," Nutritionix, accessed
August 30th, 2021, https://www.nutritionix.com/i/general-mills/
honey-nut-cheerios-cereal/51d2fb6dcc9bff111580dc91

*"When 60 million Americans eat a cereal such as"*: Statista report based on the
U.S. Census data and Simmons National Consumer Survey (NHCS).

*"the one with more carbohydrates leads to less available circulating energy"*: Kim
J. Shimy et al., "Effects of dietary carbohydrate content on circulating

metabolic fuel availability in the postprandial state," *Journal of the Endocrine Society* 4, no. 7 (2020): bvaa062, https://academic.oup.com/jes/article/4/7/bvaa062/5846215

*"A breakfast that creates a big glucose spike"*: Paula Chandler-Laney et al., "Return of hunger following a relatively high carbohydrate breakfast is associated with earlier recorded glucose peak and nadir," *Appetite* 80 (2014): 236-241, https://www.sciencedirect.com/science/article/abs/pii/S0195666314002049

*"that breakfast will deregulate our glucose levels for the rest of the day"*: Courtney R. Chang et al., "Restricting carbohydrates at breakfast is sufficient to reduce 24-hour exposure to postprandial hyperglycemia and improve glycemic variability," *American Journal of Clinical Nutrition* 109, no. 5 (2019): 1302-1309, https://academic.oup.com/ajcn/article/109/5/1302/5435774?login=true

*"A flat breakfast, on the other hand"*: Courtney R Chang et al., "Restricting carbohydrates at breakfast is sufficient to reduce 24-hour exposure to postprandial hyperglycemia and improve glycemic variability," *American Journal of Clinical Nutrition* 109, no. 5 (2019): 1302-1309, https://academic.oup.com/ajcn/article/109/5/1302/5435774?login=true

*"A century ago, the California Fruit Growers Exchange"*: Adee Braun, "Misunderstanding Orange Juice as a Health Drink," *The Atlantic* (2014), https://www.theatlantic.com/health/archive/2014/02/misunderstanding-orange-juice-as-a-health-drink/283579/

*"By blending a piece of fruit, we pulverise the fibre"*: KeXue Zhu et al., "Effect of ultrafine grinding on hydration and antioxidant properties of wheat bran dietary fiber," *Food Research International* 43, no. 4 (2010): 943-948, https://www.sciencedirect.com/science/article/abs/pii/S0963996910000232

*"One 300ml bottle of orange juice"*: U.S. Department of Agriculture, "Tropicana Pure Premium Antioxidant Advantage No Pulp Orange Juice 59 Fluid Ounce Plastic Bottle," FoodData Central, 2019, accessed August 30th, 2019, https://fdc.nal.usda.gov/fdc-app.html#/food-details/762958/nutrients

*"that's the concentrated sugar of three whole oranges"*: U.S. Department of Agriculture, "Oranges, raw, navels," FoodData Central, 2019, accessed August 30th, 2019, https://fdc.nal.usda.gov/fdc-app.html#/food-details/746771/nutrients

*"It's the same amount of sugar as in a can of Coca-Cola"*: U.S. Department of Agriculture, "Coca-Cola Life Can, 12 fl oz," FoodData Central, 2019, accessed August 30th, 2019, https://fdc.nal.usda.gov/fdc-app.html#/food-details/771674/nutrients

*"With just 300ml of orange juice,"*: American Heart Association, "Added Sugars", Heart, accessed August 30th, 2019, https://www.heart.org/en/healthy-living/healthy-eating/eat-smart/sugar/added-sugars

*"And the answer to whether sugar makes your brain work better is… no"*: Rachel

Galioto et al., "The effects of breakfast and breakfast composition on cognition in adults," *Advances in Nutrition* 7, no. 3 (2016): 576S-589S, https://academic.oup.com/advances/article/7/3/576S/4558060

*"Research shows that when diabetics replace their oatmeal"*: Martha Nydia Ballesteros et al., "One egg per day improves inflammation when compared to an oatmeal-based breakfast without increasing other cardiometabolic risk factors in diabetic patients," *Nutrients* 7, no. 5 (2015): 3449-3463, https://www.mdpi.com/2072-6643/7/5/3449

## Hack 5: Have any type of sugar you like – they're all the same

*"And misinformation is rampant"*: Republic of the Philippines Department of Science and Technology, "Glycemic Index of Coco Sugar," Internet Archive, accessed August 30th, 2019, https://web.archive.org/web/20131208042347/http://www.pca.da.gov.ph/pdf/glycemic.pdf.

*"and that was later proven to be wrong."*: University of Sydney Glycemic Index Research Service, "Glycemic Index of Coconut Sugar", Glycemic Index, accessed August 30th, 2021, https://glycemicindex.com/foodSearch.php?num=2659&ak=detail

*"fructose is worse for us than glucose"*: Robert H. Lustig, "Fructose: it's 'alcohol without the buzz'," *Advances in nutrition* 4, no. 2 (2013): 226-235, https://www.ncbi.nlm.nih.gov/pmc/articles/PMC3649103/

*"And fun fact: there aren't that many antioxidants in honey"*: There are 5.15 mg/kg of flavonoids antioxidants in multi-floral honey. One teaspoon is 4 grams. That gives 0.02mg of flavonoids per teaspoon of honey. Goran Šarić et al., "The changes of flavonoids in honey during storage." *Processes* 8, no. 8 (2020): 943, https://www.mdpi.com/2227-9717/8/8/943/pdf

100 grams of blueberries contains on average 4mg of flavonoids. One blueberry is about 1 gram. That's 0.04 mg per blueberry. De Pascual-Teresa et al., "Flavanols and anthocyanins in cardiovascular health: a review of current evidence," *International Journal of Molecular Sciences* 11, no. 4 (2010): 1679-1703, https://www.rcscarchgatc.nct/publication/44609005_Flavanols_and_Anthocyanins_in_Cardiovascular_Health_A_Review_of_Current_Evidence

*"when people switch from drinking diet drinks"*: A. Madjd et al., "Effects of replacing diet beverages with water on weight loss and weight maintenance: 18-month follow-up, randomized clinical trial," *International Journal of Obesity* 42, no. 4 (2018): 835-840, https://www.nature.com/articles/ijo2017306

*"What's more, preliminary studies suggest"*: J.E. Blundell et al., "Paradoxical effects of an intense sweetener (aspartame) on appetite," *Lancet (USA)* (1986), https://agris.fao.org/agris-search/search.do?recordID=US8731275

*"The theory posits further that"*: Susan E. Swithers et al., "A role for sweet taste: calorie predictive relations in energy regulation by rats," *Behavioral Neuroscience* 122, no. 1 (2008): 161, https://psycnet.apa.org/doiLanding?doi=10.1037%2F0735-7044.122.1.161

*"Artificial sweeteners may also change the composition of our intestinal bacteria"*:

Francisco Javier Ruiz-Ojeda et al., "Effects of sweeteners on the gut microbiota: a review of experimental studies and clinical trials," *Advances in Nutrition* 10, no. suppl_1 (2019): S31-S48, https://www.ncbi.nlm.nih. gov/pmc/articles/PMC6363527/

*"There are some artificial sweeteners I'd recommend you avoid"*: Stephen D. Anton et al., "Effects of stevia, aspartame, and sucrose on food intake, satiety, and postprandial glucose and insulin levels," *Appetite* 55, no. 1 (2010): 37-43, https://www.sciencedirect.com/science/article/abs/pii/ S0195666310000826

## Hack 6: Pick dessert over a sweet snack

*"they keep working for four hours on average after our last bite"*: Louis Monnier et al., "Target for glycemic control: concentrating on glucose," *Diabetes Care* 32, no. suppl 2 (2009): S199-S204, https://www.ncbi.nlm.nih.gov/pmc/ articles/PMC2811454/

*"The postprandial state is the period of our day"*: Maarten R. Soeters, "Food intake sequence modulates postprandial glycemia," *Clinical Nutrition* 39, no. 8 (2020): 2335-2336, https://www.clinicalnutritionjournal.com/ article/S0261-5614(20)30299-5/abstract

*"To digest, sort, and store the molecules"*: Nagham Jafar et al., "The effect of short-term hyperglycemia on the innate immune system," *The American Journal of the Medical Sciences* 351, no. 2 (2016): 201-211, https://www.amjmedsci. org/article/S0002-9629(15)00027-0/fulltext

*"Our insulin levels, oxidative stress, and inflammation increase"*: Amber M. Milan et al., "Comparisons of the postprandial inflammatory and endotoxaemic responses to mixed meals in young and older individuals: a randomised trial," *Nutrients* 9, no. 4 (2017): 354, https://www.ncbi.nlm.nih.gov/pmc/ articles/PMC5409693/

*"We tend to spend about 20 hours"*: Barry M. Popkin et al., "Does hunger and satiety drive eating anymore? Increasing eating occasions and decreasing time between eating occasions in the United States," *The American Journal of Clinical Nutrition* 91, no. 5 (2010): 1342-1347, https://academic.oup. com/ajcn/article/91/5/1342/4597335?login=true

*"up until the 1980s, people didn't snack"*: Barry M Popkin et al., "Does hunger and satiety drive eating anymore? Increasing eating occasions and decreasing time between eating occasions in the United States," *American journal of clinical nutrition* 91, no. 5 (2010): 1342-1347, https://academic. oup.com/ajcn/article/91/5/1342/4597335?login=true

*"Our organs are on clean-up duty"*: M. Ribeiro et al., "Insulin decreases autophagy and leads to cartilage degradation," *Osteoarthritis and Cartilage* 24, no. 4 (2016): 731-739, https://www.sciencedirect.com/science/article/ pii/S1063458415013709#

*"the gurgling we feel in our small intestine"*: Giulia Enders, *Gut: The Inside Story of Our Body's Most Underrated Organ* (revised edition), (Greystone Books Ltd, 2018).

*"Scientists in the Czech Republic, in 2014, tested this"*: Hana Kahleova et al.,

"Eating two larger meals a day (breakfast and lunch) is more effective than six smaller meals in a reduced-energy regimen for patients with type 2 diabetes: a randomised crossover study," *Diabetologia* 57, no. 8 (2014): 1552-1560, https://link.springer.com/article/10.1007/s00125-014-3253-5

*"the benefits are more pronounced for men"*: Leonie K. Heilbronn et al., "Glucose tolerance and skeletal muscle gene expression in response to alternate day fasting," *Obesity research* 13, no. 3 (2005): 574-581, https://pubmed.ncbi.nlm.nih.gov/15833943/

*"for women of reproductive age"*: Rima Solianik et al., "Two-day fasting evokes stress, but does not affect mood, brain activity, cognitive, psychomotor, and motor performance in overweight women," *Behavioural Brain research* 338 (2018): 166-172, https://pubmed.ncbi.nlm.nih.gov/29097329/

## Hack 7: Reach for vinegar before you eat

*"by adding vinegar before meals for three months"*: Tomoo Kondo et al., "Vinegar intake reduces body weight, body fat mass, and serum triglyceride levels in obese Japanese subjects," *Bioscience, Biotechnology, and Biochemistry* 73, no. 8 (2009): 1837-1843, https://www.tandfonline.com/doi/pdf/10.1271/bbb.90231

*"by adding vinegar before meals for three months"*: Heitor O. Santos et al., "Vinegar (acetic acid) intake on glucose metabolism: A narrative review," *Clinical Nutrition ESPEN* 32 (2019): 1-7, https://www.researchgate.net/publication/333526775_Vinegar_acetic_acid_intake_on_glucose_metabolism_A_narrative_review

*"In one study, both groups were put on a strict weight loss diet"*: Solaleh Sadat Khezri et al., "Beneficial effects of Apple Cider Vinegar on weight management, Visceral Adiposity Index and lipid profile in overweight or obese subjects receiving restricted calorie diet: A randomized clinical trial," *Journal of Functional Foods* 43 (2018): 95-102, https://www.sciencedirect.com/science/article/abs/pii/S1756464618300483

*"A Brazilian research team explained that"*: Heitor O. Santos et al., "Vinegar (acetic acid) intake on glucose metabolism: A narrative review," *Clinical Nutrition ESPEN* 32 (2019): 1-7, https://www.researchgate.net/publication/333526775_Vinegar_acetic_acid_intake_on_glucose_metabolism_A_narrative_review

*"In nondiabetics, insulin-resistant type 1 diabetics and type 2 diabetics alike"*: Farideh Shishehbor et al., "Vinegar consumption can attenuate postprandial glucose and insulin responses; a systematic review and meta-analysis of clinical trials," *Diabetes Research and Clinical Practice* 127 (2017): 1-9, https://www.researchgate.net/publication/314200733_Vinegar_consumption_can_attenuate_postprandial_glucose_and_insulin_responses_a_systematic_review_and_meta-analysis_of_clinical_trials

*"In nondiabetics, insulin-resistant type 1 diabetics and type 2 diabetics alike"*: Heitor O. Santos et al., "Vinegar (acetic acid) intake on glucose metabolism: A narrative review," *Clinical Nutrition ESPEN* 32 (2019): 1-7,

https://www.researchgate.net/publication/333526775_Vinegar_acetic_acid_intake_on_glucose_metabolism_A_narrative_review

*"The effects are also seen in women with PCOS"*: Di Wu et al., "Intake of vinegar beverage is associated with restoration of ovulatory function in women with polycystic ovary syndrome," *The Tohoku Journal of Experimental Medicine* 230, no. 1 (2013): 17-23, https://www.jstage.jst.go.jp/article/tjem/230/1/230_17/_article/-char/ja/

*"the amount of insulin also decreases"*: Panayota Mitrou et al., "Vinegar consumption increases insulin-stimulated glucose uptake by the forearm muscle in humans with type 2 diabetes," *Journal of Diabetes Research* 2015 (2015), https://www.hindawi.com/journals/jdr/2015/175204/

*"Scientists have found that the acetic acid"*: Heitor O. Santos et al., "Vinegar (acetic acid) intake on glucose metabolism: A narrative review," *Clinical Nutrition ESPEN* 32 (2019): 1-7, https://www.researchgate.net/publication/333526775_Vinegar_acetic_acid_intake_on_glucose_metabolism_A_narrative_review

*"Second, once acetic acid gets into the bloodstream"*: Heitor O. Santos et al., "Vinegar (acetic acid) intake on glucose metabolism: A narrative review," *Clinical Nutrition ESPEN* 32 (2019): 1-7, https://www.researchgate.net/publication/333526775_Vinegar_acetic_acid_intake_on_glucose_metabolism_A_narrative_review

*"It tells our DNA to reprogramme"*: Heitor O. Santos et al., "Vinegar (acetic acid) intake on glucose metabolism: A narrative review," *Clinical nutrition ESPEN* 32 (2019): 1-7, https://www.researchgate.net/publication/333526775_Vinegar_acetic_acid_intake_on_glucose_metabolism_A_narrative_review

*"Reach for vinegar first"*: Elin Östman et al., "Vinegar supplementation lowers glucose and insulin responses and increases satiety after a bread meal in healthy subjects," *European Journal of Clinical Nutrition* 59, no. 9 (2005): 983-988, https://www.nature.com/articles/1602197/

*"In the first ever study looking at vinegar"*: F Brighenti et al., "Effect of neutralized and native vinegar on blood glucose and acetate responses to a mixed meal in healthy subjects," *European Journal of Clinical Nutrition* 49, no. 4 (1995): 242-247, https://pubmed.ncbi.nlm.nih.gov/7796781/

*"Vinegar to curb a glucose spike is most useful"*: Stavros Liatis et al., "Vinegar reduces postprandial hyperglycaemia in patients with type II diabetes when added to a high, but not to a low, glycaemic index meal," *European Journal of Clinical Nutrition* 64, no. 7 (2010): 727-732, https://www.nature.com/articles/ejcn201089

*"no studies have been done to measure the effects"*: Heitor O. Santos et al., "Vinegar (acetic acid) intake on glucose metabolism: A narrative review," *Clinical Nutrition ESPEN* 32 (2019): 1-7, https://www.researchgate.net/publication/333526775_Vinegar_acetic_acid_intake_on_glucose_metabolism_A_narrative_review

*"Vinegar does not appear to damage the stomach lining"*: Heitor O. Santos et al., "Vinegar (acetic acid) intake on glucose metabolism: A narrative review,"

*Clinical Nutrition ESPEN* 32 (2019): 1-7, https://www.researchgate. net/publication/333526775_Vinegar_acetic_acid_intake_on_glucose_ metabolism_A_narrative_review

*"A 29-year old woman who consumed 16 tablespoons of ACV"*: Heitor O. Santos et al., "Vinegar (acetic acid) intake on glucose metabolism: A narrative review," *Clinical Nutrition ESPEN* 32 (2019): 1-7, https://www. researchgate.net/publication/333526775_Vinegar_acetic_acid_intake_on_ glucose_metabolism_A_narrative_review

*"Drinking it after eating"*: Tomoo Kondo et al., "Vinegar intake reduces body weight, body fat mass, and serum triglyceride levels in obese Japanese subjects," *Bioscience, biotechnology, and biochemistry* 73, no. 8 (2009): 1837- 1843, https://www.tandfonline.com/doi/pdf/10.1271/bbb.90231

*"When it comes to vinegar pills"*: Carol S. Johnston et al., "Examination of the antiglycemic properties of vinegar in healthy adults," *Annals of nutrition and metabolism* 56, no. 1 (2010): 74-79, https://www.karger.com/Article/ Abstract/272133

*"When it comes to vinegar pills"*: Carol S. Johnston et al., "Preliminary evidence that regular vinegar ingestion favorably influences hemoglobin A1c values in individuals with type 2 diabetes mellitus," *Diabetes research and clinical practice* 84, no. 2 (2009): e15-e17, https://www.sciencedirect.com/science/ article/abs/pii/S0168822709000813

## Hack 8: After you eat, move

*"The more and the harder a muscle"*: Erik A. Richter et al., "Exercise, GLUT4, and skeletal muscle glucose uptake," *Physiological reviews* (2013), https:// journals.physiology.org/doi/full/10.1152/physrev.00038.2012?view=long& pmid=23899560

*"It can increase 1,000-fold"*: Julien S. Baker et al., "Interaction among skeletal muscle metabolic energy systems during intense exercise," *Journal of Nutrition and Metabolism* 2010 (2010), https://www.hindawi.com/ journals/jnme/2010/905612/

*"Resistance exercise (weight lifting)"*: Andrew Borror et al., "The effects of postprandial exercise on glucose control in individuals with type 2 diabetes: a systematic review," *Sports Medicine* 48, no. 6 (2018): 1479-1491, https://link.springer.com/article/10.1007/s40279-018-0864-x

*"if our muscles are currently contracting"*: G. Messina et al., "Exercise causes muscle GLUT4 translocation in an insulin," *Biology and Medicine* 1 (2015): 1-4, https://www.researchgate.net/profile/ Fiorenzo_Moscatelli/publication/281774994_Exercise_Causes_ Muscle_GLUT4_Translocation_in_an_Insulin-Independent_Manner/ links/55f7e0ee08aec948c474b805/Exercise-Causes-Muscle-GLUT4- Translocation-in-an-Insulin-Independent-Manner.pdf

*"if our muscles are currently contracting"*: Stephney Whillier, "Exercise and insulin resistance," *Advances in Experimental Medicine & Biology* 1228 (2020): 137-150, https://link.springer.com/ chapter/10.1007/978-981-15-1792-1_9

*"And the longer we work out"*: Jason M.R. Gill., "Moderate exercise and post-prandial metabolism: issues of dose-response," *Journal of sports sciences* 20, no. 12 (2002): 961-967, https://shapeamerica.tandfonline.com/doi/abs/10.1080/026404102321011715

*"many different scenarios have been tested"*: Sheri R. Colberg et al., "Postprandial walking is better for lowering the glycemic effect of dinner than pre-dinner exercise in type 2 diabetic individuals," *Journal of the American Medical Directors Association* 10, no. 6 (2009): 394-397, https://www.sciencedirect.com/science/article/abs/pii/S152586100900111X

*"In a study of resistance training"*: Timothy D. Heden, "Postdinner resistance exercise improves postprandial risk factors more effectively than predinner resistance exercise in patients with type 2 diabetes," *Journal of Applied Physiology* 118, no. 5 (2015): 624-634, https://journals.physiology.org/doi/full/10.1152/japplphysiol.00917.2014

*"Among other things, it helps"*: Timothy D. Heden, "Postdinner resistance exercise improves postprandial risk factors more effectively than predinner resistance exercise in patients with type 2 diabetes," *Journal of Applied Physiology* 118, no. 5 (2015): 624-634, https://journals.physiology.org/doi/full/10.1152/japplphysiol.00917.2014

*"and reduces inflammation"*: Sechang Oh et al., "Exercise reduces inflammation and oxidative stress in obesity-related liver diseases," *Medicine and Science in Sports and Exercise* 45, no. 12 (2013): 2214-2222, https://pubmed.ncbi.nlm.nih.gov/23698242/

*"it'll be more impactful after meals"*: Andrew N. Reynolds et al., "Advice to walk after meals is more effective for lowering postprandial glycaemia in type 2 diabetes mellitus than advice that does not specify timing: a randomised crossover study," *Diabetologia* 59, no. 12 (2016): 2572-2578, https://link.springer.com/article/10.1007/s00125-016-4085-2

*"the net effect of exercise is to reduce oxidative stress"*: Goto, Sataro, Hisashi Naito, Takao Kaneko, Hae Young Chung, and Zsolt Radak. "Hormetic effects of regular exercise in aging: correlation with oxidative stress." *Applied Physiology, Nutrition, and Metabolism* 32, no. 5 (2007): 948-953, https://cdnsciencepub.com/doi/abs/10.1139/H07-092

### Hack 9: If you have to snack, go savoury

*"that in people with this mental health condition"*: Daphne Simeon et al., "Feeling unreal: a PET study of depersonalization disorder," *American Journal of Psychiatry* 157, no. 11 (2000): 1782-1788, https://ajp.psychiatryonline.org/doi/full/10.1176/appi.ajp.157.11.1782

*"Science tells us that when people eat"*: Kara L. Breymeyer et al., "Subjective mood and energy levels of healthy weight and overweight/obese healthy adults on high-and low-glycemic load experimental diets," *Appetite* 107 (2016): 253-259, https://pubmed.ncbi.nlm.nih.gov/27507131/

*"Science tells us that when people eat"*: Rachel A. Cheatham et al., "Long-term effects of provided low and high glycemic load low energy diets on mood and cognition," *Physiology & behavior* 98, no. 3 (2009): 374-379, https://

pubmed.ncbi.nlm.nih.gov/19576915/

*"Science tells us that when people eat"*: Kara L. Breymeyer et al., "Subjective mood and energy levels of healthy weight and overweight/obese healthy adults on high-and low-glycemic load experimental diets," *Appetite* 107 (2016): 253-259, https://pubmed.ncbi.nlm.nih.gov/27507131/

*"Science tells us that when people eat"*: Sue Penckofer et al., "Does glycemic variability impact mood and quality of life?," *Diabetes technology & therapeutics* 14, no. 4 (2012): 303-310, https://www.ncbi.nlm.nih.gov/pmc/articles/PMC3317401/

*"the glucose in sweets or a granola bar tends to go to storage"*: Kim J Shimy et al., "Effects of dietary carbohydrate content on circulating metabolic fuel availability in the postprandial state," *Journal of the Endocrine Society* 4, no. 7 (2020): bvaa062, https://academic.oup.com/jes/article/4/7/bvaa062/5846215

## Hack 10: Put some clothes on your carbs

*"When you do enjoy carbs"*: Lorenzo Nesti et al., "Impact of nutrient type and sequence on glucose tolerance: Physiological insights and therapeutic implications," *Frontiers in Endocrinology* 10 (2019): 144, https://www.frontiersin.org/articles/10.3389/fendo.2019.00144/full#B58

*"Even savoury snacks"*: Lesley N. Lilly et al., "The effect of added peanut butter on the glycemic response to a high–Glycemic index meal: A pilot study," *Journal of the American College of Nutrition* 38, no. 4 (2019): 351-357, https://pubmed.ncbi.nlm.nih.gov/30395790/

*"Even savoury snacks"*: David J.A. Jenkins et al., "Almonds decrease postprandial glycemia, insulinemia, and oxidative damage in healthy individuals," *Journal of Nutrition* 136, no. 12 (2006): 2987-2992, https://academic.oup.com/jn/article/136/12/2987/4663963

*"But the most recent science"*: Lorenzo Nesti et al., "Impact of nutrient type and sequence on glucose tolerance: Physiological insights and therapeutic implications," *Frontiers in Endocrinology* 10 (2019): 144, https://www.frontiersin.org/articles/10.3389/fendo.2019.00144/full#B58

*"Adding fat to a meal does not increase the insulin spike"*: Diana Gentilcore et al., "Effects of fat on gastric emptying of and the glycemic, insulin, and incretin responses to a carbohydrate meal in type 2 diabetes," *Journal of Clinical Endocrinology & Metabolism* 91, no. 6 (2006): 2062-2067, https://pubmed.ncbi.nlm.nih.gov/16537685/

*"Eating carbohydrates alone isn't just"*: Karen E Foster-Schubert et al., "Acyl and total ghrelin are suppressed strongly by ingested proteins, weakly by lipids, and biphasically by carbohydrates," *Journal of Clinical Endocrinology & Metabolism* 93, no. 5 (2008): 1971-1979, https://www.ncbi.nlm.nih.gov/pmc/articles/PMC2386677/

*"When we eat carbohydrates on their own"*: Adapted from Karen E. Foster-Schubert et al., "Acyl and total ghrelin are suppressed strongly by ingested proteins, weakly by lipids, and biphasically by carbohydrates," *Journal of Clinical Endocrinology & Metabolism* 93, no. 5 (2008): 1971-1979, https://

www.ncbi.nlm.nih.gov/pmc/articles/PMC2386677/

*"But thanks to a fascinating recent experiment"*: Sabrina Strang et al., "Impact of nutrition on social decision making," *Proceedings of the National Academy of Sciences* 114, no. 25 (2017): 6510-6514, https://www.pnas.org/content/114/25/6510/

*"Remember: when we combine glucose"*: Lorenzo Nesti et al., "Impact of nutrient type and sequence on glucose tolerance: Physiological insights and therapeutic implications," *Frontiers in Endocrinology* 10 (2019): 144, https://www.frontiersin.org/articles/10.3389/fendo.2019.00144/full

**You are special**

*"Starting in 2015, research teams"*: Sarah E. Berry et al., "Human postprandial responses to food and potential for precision nutrition," *Nature Medicine* 26, no. 6 (2020): 964-973, https://www.nature.com/articles/s41591-020-0934-0

*"Some studies even found that if you think"*: Chanmo Park et al., "Glucose metabolism responds to perceived sugar intake more than actual sugar intake," *Scientific Reports* 10, no. 1 (2020): 1-8, https://www.nature.com/articles/s41598-020-72501-w

# Index

**JESSIE INCHAUSPÉ** is a French biochemist and
**#1 *Sunday Times* bestselling author.** She is on a mission to
translate cutting-edge science into easy tips to help people
improve their physical and mental health. In her books
*Glucose Revolution* and *The Glucose Goddess Method*, she
shares her startling discovery about the essential role of blood
sugar in every aspect of our lives, and the surprising hacks to
optimise it. Jessie is the founder of the wildly popular
Instagram account *@GlucoseGoddess*, where she teaches
over two million people about transformative food habits.
She holds a BSc in mathematics from King's College,
London, and an MSc in biochemistry from
Georgetown University.

Sign up to Jessie's newsletter to stay up to
date on all the latest glucose discoveries.